TOF Range-Imaging Cameras

Fabio Remondino · David Stoppa
Editors

TOF Range-Imaging Cameras

Springer

Editors
Fabio Remondino
3D Optical Metrology
Fondazione Bruno Kessler
Trento
Italy

David Stoppa
Smart Optical Sensors and Interfaces
Fondazione Bruno Kessler
Trento
Italy

ISBN 978-3-642-27522-7 ISBN 978-3-642-27523-4 (eBook)
DOI 10.1007/978-3-642-27523-4
Springer Heidelberg New York Dordrecht London

Library of Congress Control Number: 2013935255

Printed on acid-free paper

Springer is part of Springer Science+Business Media (www.springer.com)

Contents

State-of-the-Art of TOF Range-Imaging Sensors

Dario Piatti, Fabio Remondino and David Stoppa

1 Introduction

The 3D information of a surveyed object or scene can be recorded with different types of sensors and measuring techniques. Contactless measuring techniques suitable to estimate the target distance exploit micro-, ultrasonic- or light- waves [1, 2]. However only the latter technique allows achieving good angular resolution performance, in a compact measuring setup, as required for a 3D imaging system [3]. In the common practice, the two ways to acquire an object's geometry are: (i) passive, by using multi-view image data or (ii) active, exploiting optical distance measurement techniques.

The multi-view image acquisition method, coupled with the triangulation measurement principle, is already known and used for decades in the research community [4]. One of the advantages of the image approach with respect to other range measuring devices (such as LiDAR, acoustic or radar sensors) is the reachable high resolution and simultaneous acquisition of the surveyed area without energy emission or moving parts. Still, the major disadvantages are the correspondence problem, the processing time and the need of adequate illumination conditions and textured surfaces in the case of automatic matching procedures.

Active optical measuring techniques using light-waves can be further classified in three main categories, namely: interferometry, triangulation and Time-Of-Flight (TOF) [5–7]. Triangulation techniques normally determines an unknown point

D. Piatti (✉)
Politecnico di Torino, Corso Duca degli Abruzzi 24 10129 Turin, Italy
e-mail: dario.piatti@polito.it

F. Remondino · D. Stoppa
Fondazione Bruno Kessler, Via Sommarive 18 38123 Trento, Italy
e-mail: remondino@fbk.eu

D. Stoppa
e-mail: stoppa@fbk.eu

F. Remondino and D. Stoppa (eds.), *TOF Range-Imaging Cameras*,
DOI: 10.1007/978-3-642-27523-4_1,

within a triangle by means of a known optical basis and the related side angles pointing to the unknown point. This principle is used by active sensors based on structured illumination as well as by passive digital cameras.

Continuous wave and pulse TOF techniques measure the time of flight or the phase shift of a modulated optical signal. These techniques usually apply incoherent optical signals. Typical examples of TOF are the optical rangefinder of total stations or classical LiDAR instruments (terrestrial or aerial) [8, 9]. In this latter case, actual laser scanners allow to acquire almost one million of points per second, thanks to fast scanning mechanisms. Their measurement range can vary to a great extent according to the instruments, varying between some decimeters up to some kilometers, with an accuracy ranging from less than one millimeter to some tens of centimeters respectively. Nevertheless, the main drawbacks of LiDAR instruments are their high costs and dimensions.

Interferometry methods measure depths by means of the Time-Of-Flight techniques too. In this case, however, the phase of the optical wave itself is used. This requires coherent mixing and correlation of the wave-front reflected from the object with a reference wave-front. Many variants of the optical interferometry principle have been developed, such as multi-wavelength interferometry, holographic interferometry, speckle interferometry and white light interferometry. The high accuracy of the interferometry methods mainly depend on the coherence length of the light source: interferometry is not suitable for ranges greater than few centimeters since the method is based on the evaluation of very short optical wavelength.

In the last few years a new generation of active sensors has been developed, which allows to acquire 3D point clouds without any scanning mechanism and from just one point of view at video frame rates. The working principle is the measurement of the TOF of an emitted signal by the device towards the object to be observed, with the advantage of simultaneously measuring the distance information for each pixel of the camera sensor. Many terms have been used in the literature to indicate such devices, normally called Time-Of-Flight (TOF) cameras, Range IMaging (RIM) cameras, 3D range imagers, range cameras or a combination of these terms. In the following sections and chapters the term TOF cameras will be prevalently employed, which is more related to the working principle of this technology. Such a technology is possible because of the miniaturization of the semiconductor technology and the evolvement of the CCD/CMOS processes that can be implemented independently for each pixel. Thus it is possible to acquire distance measurements for each pixel at high frame rate and with accuracies up to few centimeters. While TOF cameras based on the phase-shift measurement usually have a working range limited to ten/thirty meters, TOF cameras based on the direct TOF measurement can measure distances up to 1500 m. Moreover, TOF cameras are usually characterized by low resolution (no more than a few thousands of tens of pixels), small dimensions, costs that are an order of magnitude lower with respect to LiDAR instruments and a lower power consumption with respect to classical laser scanners. In contrast to multi-view image acquisitions, the depth accuracy is practically independent of textural appearance, but limited to about one centimeter in the best case.

Recently a great alternative to TOF cameras came on the market: it is the line of sensors based on real-time pattern projection and triangulation technique which enable simultaneous acquisition of geometry and texture, at low-cost, high frame rate and with ranges up to 4–5 m. The most well know sensor of this family is the Microsoft Kinect [10, 11]. This book will not touch such devices as they are not based on the TOF measurement principle.

In order to give an overview on the TOF cameras technology, this chapter will provide a quick introduction of the TOF cameras operation principle and a description of their main building blocks. Then, the main technologies available today for the realization of TOF detectors will be described and compared and finally some conclusions and future perspective will be given.

2 Working Principle of TOF Cameras

2.1 TOF Detection System

A typical TOF measuring setup is sketched in Fig. 1, and it consists of several building blocks: (a) a pulsed/modulated light source, typically based on LASER or LED in the infrared part of the spectrum to make the illumination unobtrusive, (b) an optical diffuser to spread the emitted light onto the scene, (c) a collection lens aimed at gathering the light echo back-reflected by the target. An optical band-pass, properly tuned onto the wavelength of the light source, allows improving the background noise rejection. Finally, the core of the measuring system is represented by the solid-state range sensor (d), composed of an array of photo-detectors (pixels) capable of measuring, in a direct on indirect way, the TOF needed by the light pulse to travel from the light source to the target and back to the sensor. The system requires also a suitable sensor interface providing to the sensor the power supply, required biasing voltage/current signals, digital control phases, and reading out from the sensor the data stream, which typically requires further minor processing to obtain the 3D volume data. Finally, the sensor interface is responsible for the communication with the external world (to a PC, or a processing unit).

2.2 TOF Measurement Techniques

In a classical TOF measurement, referred in the following as Direct-TOF (D-TOF), the detector system starts a highly accurate stopwatch synchronously with the emitter light pulse generation. As the light echo from the target is detected, the stopwatch is stopped and the roundtrip time τ_{TOF} is directly stored. The target distance z can be estimated by means of the simple equation:

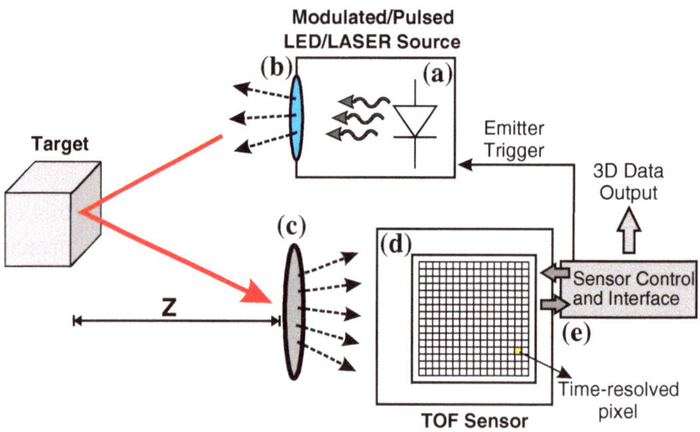

Fig. 1 TOF detection system

$$Z = \frac{c}{2} \cdot \tau_{TOF} \tag{1}$$

where c = 2.9979×10^8 m/s represents the speed of the light propagating through the air.

D-TOF is commonly used for single-point range systems, but only recently implemented in scannerless TOF systems, because of the difficulties in implementing at pixel level sub-nanosecond electronic stopwatch. This technique is particularly suited to SPAD-based TOF systems [12–23] and details about its implementation will be described in Chap. 2.

An alternative solution to D-TOF is the so-called Indirect-TOF (I-TOF), where the roundtrip trip time is indirectly extrapolated from a time-gated measurement of the light intensity. In this case, there is no need of precise stopwatch, but of time-gated photons counters or charge integrators, which can be implemented at a pixel level with less efforts and silicon area. I-TOF is the natural solution for electronic- and photo-mixing devices-based TOF cameras.

The operation principle of D-TOF and an example of a four-gates I-TOF are illustrated in Fig. 2 considering both pulsed and modulated light sources, although many other implementations of I-TOF are possible. I-TOF will be extensively described in Chaps. 3 and 4 together with its circuital implementation.

3 Time-Resolved Image Sensor Technologies

Although there are many TOF systems based on laser scanner available on the market for top-class 3D measurement apparatus, there has been in the last decade an emerging interest toward scannerless, all-solid-state, TOF cameras. Many

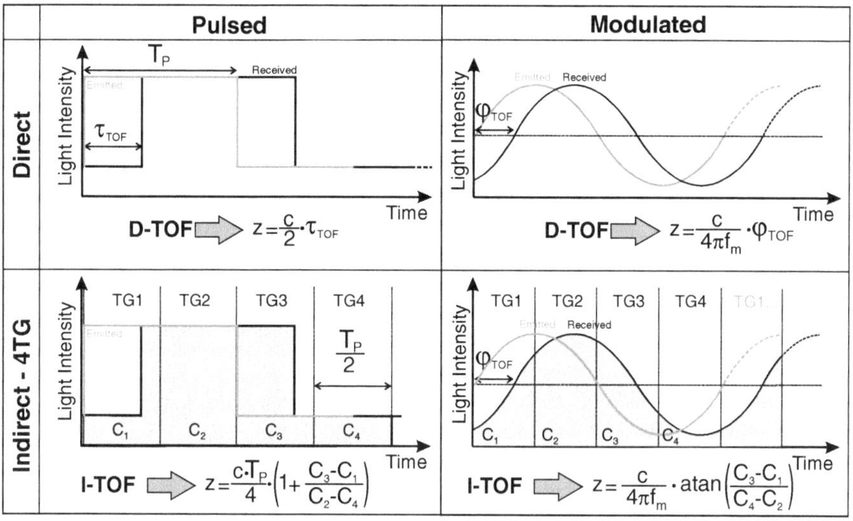

Fig. 2 Overview of pulsed and modulated D-TOF and I-TOF measuring techniques

full-custom detectors have been developed for this class of systems. They can be mainly classified in three categories:

(a) *In-Pixel Photo-mixing Devices*. This approach exploits photo-demodulators, in which the photo-generated charge is mixed toward two or more collection electrodes thus achieving an intrinsic photo-mixing effect.
(b) *Standard photodiodes coupled to dedicated processing circuitry*. This approach exploits the extensive use of switched-capacitors electronics (either in the pixel or at the periphery) to recover the distance information from the current photo-generated by the photodiode.
(c) *Single-Photon Avalanche Diodes (SPADs) coupled to proper processing circuitry*. With respect to (b), an avalanche diode, operating in Geiger regime to achieve sensitivity to individual photons, is used to collect the light echo and coupled to a readout and processing electronics aimed at extracting the Time-Of-Flight information.

The most mature solution is represented by sensors belonging to a), and most of the 3D cameras available on the market are actually based on this concept [24–39]. The main advantage of this approach is the read-out channel simplicity, which results in a small pixel size; the main problems are the sensitivity to the ambient light and the cost of non-standard technologies (e.g. CCD/CMOS, customized CMOS, high resistivity substrate, etc.), that are often required.

The use of complex in-pixel electronics, used to properly process and accumulate the photo-generated charge, makes sensors belonging to b) being characterized by large pixel pitch and relatively high power consumption [40–48]. Moreover, they typically exhibit lower precision with respect to (a) and (c) because of the noise contribution introduced by the numerous transistors introduced in the signal path. On the other hand, ad hoc processing structures can be implemented at pixel- or column-level to successfully remove most of the common mode signal, thus implementing background resilient sensors that can be used for outdoor operation such as automotive and security.

Finally, sensors based on category (c) have been widely used in high-performance single-point scanner systems since the 80s, using avalanche photodiodes (APDs) fabricated within dedicated technologies coupled to discrete-components read out and processing electronics or measurement instruments. However, only recently, the possibility of implementing good performance APDs/SPADs in CMOS technologies opened the way to the realization of range image sensors based on this approach [12–23, 49–51].

The great advantage of this solution is the extremely high sensitivity of the photodetector, capable of detecting down to single-photon, and the intrinsic low-noise performance that allows operating at the shot-noise limit.

The above mentioned sensor categories will be extensively described in Chaps. 2, 3 and 4 where the state-of-the-art for each approach will be analyzed through circuital and device implementations peculiar of each sensor architecture.

4 Conclusions

Image sensors capable of detecting arrival times of impinging light signals with sub-nanosecond time resolution are becoming more and more important in many applications. Among them, TOF 3D cameras represent one of the markets having the largest possibilities of expansion thanks to the numerous sectors of exploitation of such a technology. The extra information provided by 3D cameras, with respect to standard 2D imagers, is fundamental to acquire a reliable model of the scene under measurement, opening the way to new detection paradigms in the field of machine vision. Industrial control, next generation user interfaces based on gesture recognition, advanced vision systems for automotive, etc. are just a few examples of important sectors using 3D-imaging technology.

In this chapter the operation principle of TOF cameras has been described and the main TOF sensor architectures presented so far have been reviewed. This is mainly an introduction to extended descriptions provided in Chaps. 2, 3 and 4 that will deal with TOF sensors based on SPAD, electronic shutter and photo-mixing devices respectively.

Regardless of the numerous solutions proposed in the scientific literature in the last few years and although several commercial products are now available in the field of TOF 3D cameras, there are still several aspects that can be improved. Ambient light immunity and dynamic range enhancement are two key features for the next generation of 3D cameras, while as in conventional cameras, there is a continuous demand of higher resolution and frame rate, as well as reduced power consumption. The addition of color in the same sensor is also a very appealing feature, and the first attempts to implement this functionality are being carried out.

The development of TOF technology should also take into account competing technologies, such as those based on pattern projection and stereo imaging. Those systems successfully demonstrated that accessing consumer market could dramatically reduce the system cost. In fact, the building blocks of system like [11] are quite similar to the ones needed by TOF cameras, i.e. illuminator, custom image sensor, and optics, however, the final cost of this system is one order of magnitude lower than the cost of TOF camera products.

The main drawbacks of systems based on pattern projection and stereo image matching with respect to TOF technology are: (i) the limited scalability of the system size due to the need of a baseline (ii) the high computational effort required to extract the depth information that limits the sensor frame rate and the minimum power consumption, and finally (iii) the generation of artifacts under some measurement conditions.

The next three to five years will demonstrate all the potential of TOF technology and will reveal if the evolution of TOF cameras will follow the same amazing expansion experienced by conventional CMOS cameras in the 2000s.

References

1. R. Schwarte, Principles of 3-D Imaging Technology, in *Handbook of Computer Vision and Applications*, ed. by B. Jähne, H. Haussecker, P. Geißler (Academic Press, 1999)
2. B. Jähne, H. Haussecker, P. Geißler, *Handbook of Computer Vision and Applications*, vol. 1 (Academic Press, San Diego, 1999) pp. 479–482
3. P. Besl, Active optical range imaging sensors. Mach. Vis. Appl. **1**, 127–152 (1988)
4. E.M. Mikhail, J.S. Betherl, J.C. McGlone, *Introduction to modern photogrammetry* (John Wiley & Sons, Inc., New York, 2001)
5. M.-C. Amann, T.B.M. Lescure, R. Myllyla, M. Rioux, Laser ranging: a critical review of usual techniques for distance measurement. Opt. Eng. **40**, 10–19 (2001)
6. F. Blais, Review of 20 years of range sensor development. J. Elect. Imaging **13**(1), 231–243 (2004)
7. B. Hosticka, P. Seitz, A. Simoni, Optical Time-Of-Fight sensors for solid state 3D-vision. Encycl. Sens. **7**, 259–289 (2006)
8. Leica-geosystem website, www.leica-geosystems.com
9. Konicaminolta website, www.konicaminolta.com
10. F. Menna, F. Remondino, R. Battisti, E. Nocerino, Geometric investigation of a gaming active device. Proc. SPIE Opt. Metrol. **8085**(1), 80850G (2011)
11. K. Khoshelham, sO Elberink, Accuracy and resolution of kinetic depth data for indoor mapping applications. Sensors **12**, 1437–1454 (2012). doi:10.3390/s120201437
12. M.A. Albota et al. Three-dimensional imaging laser radars with geiger-mode avalanche photodiode arrays. Lincoln Labs J. **13**(2) 351–370 (2002)
13. C. Niclass, A. Rochas, P.A. Besse, E. Charbon, Design and characterization of a CMOS 3-D image sensor based on single photon avalanche diodes. IEEE J. Solid-State Circuits **40**(9), 1847–1854 (2005)

14. J. Richardson, R. Walker, L. Grant, D. Stoppa, F. Borghetti, E. Charbon, M. Gersbach, R. K. Henderson, A 32 × 32 50 ps resolution 10 bit time to digital converter array in 130 nm CMOS for time correlated imaging, in *IEEE Custom Integrated Circuits Conference* (2009) pp. 77–80
15. M. Gersbach, Y. Maruyama, E. Labonne, J. Richardson, R. Walker, L. Grant, R. K. Henderson, F. Borghetti, D. Stoppa, E. Charbon, "A Parallel 32 × 32 time-to-digital converter array fabricated in a 130 nm imaging CMOS technology", in *IEEE European Solid-State Device Conference* (2009)
16. C. Veerappan, J. Richardson, R. Walker, D.U. Li, M.W. Fishburn, Y. Maruyama, D. Stoppa, F. Borghetti, M. Gersbach, R.K. Henderson, E. Charbon, A 160 × 128 single-photon image sensor with on-pixel, 55 ps 10b time-to-digital converter, in *IEEE International Solid-State Circuits Conference* (2011) pp. 312-314
17. D. Stoppa, F. Borghetti, J. Richardson, R. Walker, L. Grant, R.K. Henderson, M. Gersbach, E. Charbon, "A 32 × 32-pixel array with in-pixel photon counting and arrival time measurement in the analog domain, in *IEEE European Solid-State Device Conference* (2009) pp. 204–207
18. R. J. Walker, J. R. Richardson, R. K. Henderson; A 128 × 96 pixel event-driven phase-domain ΔΣ-based fully digital 3D camera in 0.13 μm CMOS imaging technology, *IEEE International Solid-State Circuits Conference* (2011) pp. 410–412
19. M. A. Itzler, M. Entwistle, M. Owens, K. Patel, X. Jiang, K. Slomkowski, S. Rangwala, Geiger-mode avalanche photodiode focal plane arrays for three-dimensional imaging LADAR. SPIE Infrared Remote Sens. Instrum. 7808 (2010)
20. B Aull, J. Burns, C. Chenson, B. Felton, H. Hanson, C. Keast, J. Knecht, A. Loomis, M. Renzi, A. Soares, S. Vyshnavi, K. Warner, D. Wolfson, D. Yost, D. Young, Laser radar imager based on 3D integration of geiger-mode avalanche photodiodes with two SOI timing circuit layers. IEEE Int. Solid-State Circuits Proc. 304–305 (2006)
21. C. Niclass, C. Favi, T. Kluter, F. Monnier, E. Charbon, Single-photon synchronous detection. IEEE J. Solid-State Circuits **44**(7), 1977–1989 (2009)
22. C. Niclass, M. Soga, S. Kato, A 0.18 μm CMOS single-photon sensor for coaxial laser rangefinders, in *Asian Solid-State Circuits Conference* (2010)
23. L. Pancheri, N. Massari, F. Borghetti, D. Stoppa, A 32 × 32 SPAD pixel array with nanosecond gating and analog readout, *International Image Sensor Workshop* (*IISW*), Hokkaido 8–11 June 2011
24. Mesa Imaging website, www.mesa-imaging.ch
25. PMD Technologies website, www.pmdtec.com
26. SoftKinetic website, www.softkinetic.com
27. T. Spirig, P. Seitz, O. Vietze, F. Heitger, The lock-in CCD two dimensional synchronous detection of light. IEEE J. Quantum Electron. **31**, 1705–1708 (1995)
28. R. Miyagawa, T. Kanade, CCD-based range-finding sensor. IEEE Trans. Electron Dev. **44**(10), 1648–1652 (1997)
29. S. Kawahito, I.A. Halin, T. Ushinaga, T. Sawada, M. Homma, Y. Maeda, A CMOS Time-Of-Flight range image sensor with gates-on-field-oxide structure. IEEE Sens. J. **7**(12), 1578–1586 (2007)
30. D. Van Nieuwenhove, W. Van Der Tempel, M. Kuijk, Novel standard CMOS detector using majority current for guiding photo-generated electrons towards detecting junctions, in *Proceedings of IEEE/LEOS Symposium, Benelux Chapter*, pp. 229–232 (2005)
31. W. van der Tempel, R. Grootjans, D. Van Nieuwenhove, M. Kuijk, A 1k-pixel 3-D CMOS sensor, in *Proceedings of IEEE Sensors Conference* (2008) pp. 1000–1003
32. G.-F. Dalla Betta, S. Donati, Q.D. Hossain, G. Martini, L. Pancheri, D. Saguatti, D. Stoppa, G. Verzellesi, Design and characterization of current-assisted photonic demodulators in 0.18-μm CMOS technology. IEEE Trans. Electron Dev. **58**(6), 1702–1709 (2011)
33. L. Pancheri, D. Stoppa, N. Massari, M. Malfatti, L. Gonzo, Q. D. Hossain, G.-F. Dalla Betta, A 120x160 pixel CMOS range image sensor based on current assisted photonic

demodulators. in *Proceedings of SPIE*, vol. 7726 (SPIE Photonics Europe, Brussels, Belgium, 2010) pp. 772615
34. D. Stoppa, N. Massari, L. Pancheri, M. Malfatti, M. Perenzoni, L. Gonzo, A range image sensor based on 10-μm lock-in pixels in 0.18-μm CMOS imaging technology. IEEE J. Solid-State Circuits **46**(1), 248–258 (2011)
35. H.-J. Yoon, S. Itoh, S. Kawahito, A CMOS image sensor with in-pixel two-stage charge transfer for fluorescence lifetime imaging. IEEE Trans. Electron Dev. **56**(2), 214–221 (2009)
36. L.-E. Bonjour, T. Baechler, M. Kayal, High-speed general purpose demodulation pixels based on buried photodiodes, in *Proceedings of IISW* 2011 (Hokkaido, June 8–11, 2011)
37. C. Tubert, L. Simony, F. Roy, A. Tournier, L. Pinzelli, P. Magnan, High speed dual port pinned-photodiode for Time-Of-Flight imaging, in *Proceedings of IISW 2009* (Bergen, Norway, June 26–28, 2009)
38. H. Takeshita, T. Sawada, T. Iida, K. Yasutomi, S. Kawahito, High-speed charge transfer pinned-photodiode for a CMOS Time-Of-Flight range image sensor. Proc. SPIE **7536**, 75360R (2010)
39. S.-J. Kim, J.D.K. Kim, S.-W. Han, B. Kang, K. Lee, C.-Y. Kim "A 640 × 480 image sensor with unified pixel architecture for 2D/3D imaging in 0.11 μm CMOS. IEEE Symp VLSI Circuits, 92–93 (2011)
40. R. Jeremias, W. Brockherde, G. Doemens, B. Hosticka, L. Listl, P. Mengel, A CMOS photosensor array for 3D imaging using pulsed laser. IEEE Int. Solid-State Circuits Conf. 252–253 (2001)
41. D. Stoppa, L. Viarani, A. Simoni, L. Gonzo, M. Malfatti and G. Pedretti, "A 50 × 30-pixel CMOS sensor for TOF-based Real Time 3D Imaging", *Workshop on Charge-Coupled Devices and Advanced Image Sensors*, Karuizawa, Nagano, 2005
42. M. Perenzoni, N. Massari, D. Stoppa, L. Pancheri, M. Malfatti, L. Gonzo, A 160 × 120-pixels range camera with in-pixel correlated double sampling and fixed-pattern noise correction. IEEE J. Solid-State Circuits **46**(7), 1672–1681 (2011)
43. O. Sgrott, D. Mosconi, M. Perenzoni, G. Pedretti, L. Gonzo, D. Stoppa, A 134-pixel CMOS sensor for combined Time-Of-Flight and optical triangulation 3-D imaging. IEEE J. Solid-State Circuits **45**(7), 1354–1364 (2010)
44. K. Oberhauser, G. Zach, H. Zimmermann, Active bridge-correlator circuit with integrated PIN photodiode for optical distance measurement applications, in *Proceedings of the 5th IASTED International Conference Circuits, Signals and Systems* (July 2007) pp. 209–214
45. G. Zach, A. Nemecek, H. Zimmermann, Smart distance measurement line sensor with background light suppression and on-chip phase generation, in *Proceedings of SPIE, Conference on Infrared Systems and Photoelectronic Technology III*, vol. 7055 (Aug 2008) pp. 70550P1–70550P10
46. G. Zach, H. Zimmermann, A 2 × 32 range-finding sensor array wit pixel-inherent suppression of ambient light up to 120klx, in *IEEE International Solid-State Circuits Conference* (2009) pp. 352–353
47. G. Zach, M. Davidovic, H. Zimmermann, A 16 × 16 pixel distance sensor with in-pixel circuitry that tolerates 150 klx of ambient light. IEEE J. Solid-State Circuits **45**(7), 1345–1353 (2010)
48. C. Niclass, C. Favi, T. Kluter, M. Gersbach, E. Charbon, A 128 × 128 single-photon image sensor with column-level 10-bit time-to-digital converter array. IEEE J. Solid-State Circuits **43**(12), 2977–2989 (2008)
49. C. Niclass, M. Sergio, E. Charbon, A CMOS 64x48 single photon avalanche diode array with event-driven readout, in *IEEE European Solid-State Circuit Conference* (2006)
50. J.S. Massa, G.S. Buller, A.C. Walker, S. Cova, M. Umasuthan, A.M. Wallace, Time-pf-flight optical ranging system based on time-correlated single-photon counting. App. Opt. **37**(31), 7298–7304 (1998)
51. D. Stoppa, L. Pancheri, M. Scandiuzzo, L. Gonzo, G.-F. Della Betta, A. Simoni, A CMOS 3-D imager based on single photon avalanche diode. IEEE Trans. Circuits Syst. **54**(1), 4–12 (2007)

SPAD-Based Sensors

Edoardo Charbon, Matt Fishburn, Richard Walker, Robert K. Henderson and Cristiano Niclass

1 Introduction

3D imaging and multi-pixel rangefinding constitute one of the most important and innovative fields of research in image sensor science and engineering in the past years. In rangefinding, one computes the Time-Of-Flight of a ray of light, generated by a mono-chromatic or wide-spectral source, from the source through the reflection of a target object and to a detector. There exist at least two techniques to measure the Time-Of-Flight (TOF): a direct and an indirect technique. In direct techniques (D-TOF), the time difference between a START pulse, synchronized with the light source, and a STOP signal generated by the detector is evaluated. In indirect techniques (I-TOF), a continuous sinusoidal light wave is emitted and the phase difference between outgoing and incoming signals is measured. From the phase difference, the time difference is derived using well-known formulae.

Single-photon avalanche diodes (SPADs) or Geiger-mode avalanche photodiodes (GAPDs) are detectors capable of capturing individual photons with very high time-of-arrival resolution, of the order of a few tens of picoseconds. They

E. Charbon (✉) · M. Fishburn
TU Delft, Mekelweg 4 2628CD Delft, The Netherlands
e-mail: e.charbon@tudelft.nl

M. Fishburn
e-mail: m.w.fishburn@tudelft.nl

R. Walker · R. K. Henderson
The University of Edinburgh, Faraday Bldg., King's Buildings
EH9 3JL Edinburgh, Scotland, U.K
e-mail: richard.walker@ed.ac.uk

R. K. Henderson
e-mail: robert.henderson@ed.ac.uk

C. Niclass
EPFL, 1015 Lausanne, Switzerland
e-mail: cristiano.niclass@epfl.ch

F. Remondino and D. Stoppa (eds.), *TOF Range-Imaging Cameras*,
DOI: 10.1007/978-3-642-27523-4_2,

may be fabricated in dedicated silicon processes or in standard CMOS technologies. Most SPADs generally operate at room temperature, but they may also be cooled for better noise performance. Even though solid-state SPADs implemented in III–V materials exist, and the literature on the subject is extensive, in this chapter, we limit our attention to silicon devices.

Cova and McIntyre started advocating the use of SPADs for fast timing applications in the 1980s [1, 2]. Thanks to their picosecond timing resolution, SPADs are a natural candidate for D-TOF techniques. If one wants a distance resolution of, say, 1 mm, one needs to discriminate light pulses with a resolution of 6.6 ps (a round-trip Time-Of-Flight is assumed). However, such a time uncertainty is not achievable with a room temperature SPAD implemented in any silicon technology. Thus, averaging and multi-measurement techniques must be employed. A common choice is the use of time-correlated single-photon counting (TCSPC). The technique assumes that a START or synchronization signal is always present at the beginning of each measurement cycle, while the STOP signal is provided by the detector, in our case a SPAD, at a much smaller frequency, typically 10^4–10^6 smaller than that of the synchronization. If this condition is satisfied—and it is often required to minimize pile-up effects—several thousands of time-of-arrival evaluations are needed for each frame to achieve an accurate Time-Of-Flight measurement. A reverse START-STOP arrangement is also possible, depending upon the architecture of the imager, whereas a higher jitter of the measurement system might incur if high intra-optical-pulses jitter is present in the light source.

SPADs may also be used in I-TOF mode of operation. In this chapter two complementary techniques are presented. The common principle is that of quickly switching the SPAD to divert its output signal appropriately, depending on the time-of-arrival. Indirect detection techniques may also be handled in a completely digital environment, thus simplifying the electronics and control signals and enabling large arrays of pixels to be implemented in standard CMOS technologies.

The chapter is organized as follows. After an introduction to the physical mechanisms underlying SPADs, an overview of SPAD fabrication techniques is given, followed by a detailed description of direct and indirect techniques for Time-Of-Flight measurement; the description is complemented by examples of SPAD array implementations in CMOS and relative characterization. An outlook concludes the chapter.

2 The Physics of SPADs

An avalanche photodiode (APD) is a p-n junction that relies on impact ionization effects to multiply photon-generated electrons and holes. APDs output a pulse of electric current synchronous, with some time uncertainty, to the arrival of a single photon. In APD-based, Time-Of-Flight detectors and image sensors, external circuitry senses and analyzes this current pulse to find the photon's time-of-arrival, thus inferring the range; this process, known as rangefinding, enables the reconstruction

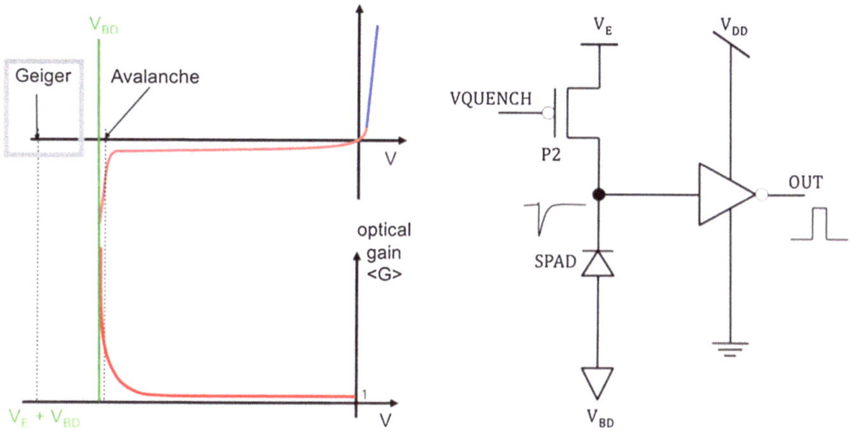

Fig. 1 Steady-state I–V characteristics for a p-n junction with Geiger and avalanche mode of operation (*left*). Passively quenched SPAD (*right*). V_E is known as the excess bias voltage at which a diode must be biased in Geiger mode; it represents the voltage in excess of or above the breakdown voltage V_{BD}. A comparator or inverter is used to shape the output pulse of the detector

of 3D scenes non-destructively. This section gives an overview of the fundamental mechanisms governing the avalanche pulse, focusing on the factors that contribute to the time uncertainty and ultimately the distance estimation accuracy.

Generally, when electrical engineers think of the I–V characteristics of a p-n junction, they think of the steady state curve, shown in Fig. 1. However, there is a pseudo-steady-state in the breakdown operating condition—a voltage above[1] the breakdown voltage can be applied so long as no carriers exist in the diode's depletion region. As soon as a carrier is injected into the depletion region, impact ionization may cause an avalanche to occur, and the diode will shift operating points to the steady-state curve. Impact ionization's underlying statistical process, which is dependent on the electric field, material, and ambient conditions, governs the probability that an avalanche will occur. If the electric field magnitude is high enough, then both electrons and holes are expected to cause significant ionization, the avalanche will become self-sustaining, and the avalanche photodiode is operating in Geiger mode.[2]

If the field magnitude is only sufficient for electrons to cause significant ionization but not holes, the APD is in linear mode. Quantitatively a diode is in Geiger mode when it meets the condition

$$1 \leq \int_0^W \alpha \cdot \exp\left(\int_x^W (\beta - \alpha) \mathrm{d}x' \right) \mathrm{d}x, \tag{1}$$

[1] The preposition "above" is used because researchers working with APDs consider the cathode to be the terminal with the higher voltage.

[2] Termed after the similarity to a Geiger counter.

with α and β representing the impact ionization rates (per m) of electrons and holes, respectively, and W is the depletion width. The bias at which Eq. (1) is an equality is called the breakdown voltage, V_{BD}. The bias applied to the diode, V_{OP}, exceeds breakdown by a voltage known as excess bias voltage, V_E. Henceforth the discussion will be restricted to SPAD technologies [3].

A SPAD is able to detect more than a single carrier by the inclusion of external circuitry, which quenches an avalanche by: sensing an avalanche; lowering the voltage applied to the SPAD; and after some time, raising the applied voltage above the breakdown voltage. The simplest such circuit is a resistor placed in series with the diode, also known as ballast resistor. The circuit works as follows. First, when there are no free carriers in the junction, the applied voltage on the diode is V_{OP}. When light injects a carrier into the junction, impact ionization may or may not cause the rapid build-up of free carriers in a small part of the diode. If significant ionization does not occur and all free carriers are swept out of the depletion region, the incident photon is not detected. If ionization does occur, it will continue until the space-charge phenomena limits the local carrier concentration. The avalanche will spread to other regions of the diode via a multiplication-assisted diffusion process. The decrease in voltage across the diode, dependent on both any parasitic capacitances and the quenching resistor, will eventually reach the excess bias. At this point the diode is quenched—no further current should flow from impact ionization and there are no free carriers in the diode itself. The voltage will then be recharged by the flow of electric current through the quenching resistor into the parasitic capacitances, with the diode being ready to detect another carrier after this dead time [4].

The probability that a single photon's generated carriers are detected is called the photon detection probability (PDP). A number of factors influence the PDP, including electric field conditions, doping levels, whether electrons or holes primarily initiate avalanches, and the applied voltage. Photons are not the only source of initial carriers; uncorrelated or correlated noise can also cause free carrier injection and undesirable avalanches. Uncorrelated noise sources include: ambient light; tunneling from an electric field that is too high; and fabrication defects which ease valence-to-conduction band transitions, such as thermally generated or tunneling carriers. The last factor, fabrication defects, can also cause afterpulsing, a type of time-correlated noise. Traps in the forbidden energy band can fill, only to release free carriers on the time scale of tens of nanoseconds following an avalanche. Afterpulsing prevents an instant recharge phase in SPADs, and places constraints on the minimum dead time.

There are other sources of correlated noise besides afterpulsing—optical and electrical crosstalk are the most common, but in practice afterpulsing is the dominant type of correlated noise. Whether the initial carrier is photon-generated or not, several factors cause time uncertainty between the injection time and the sense time. The most important timing factor is whether the carrier is generated in the depletion region itself, or if it must diffuse into the depletion region. Carriers that successfully diffuse into the depletion region do so following an exponential

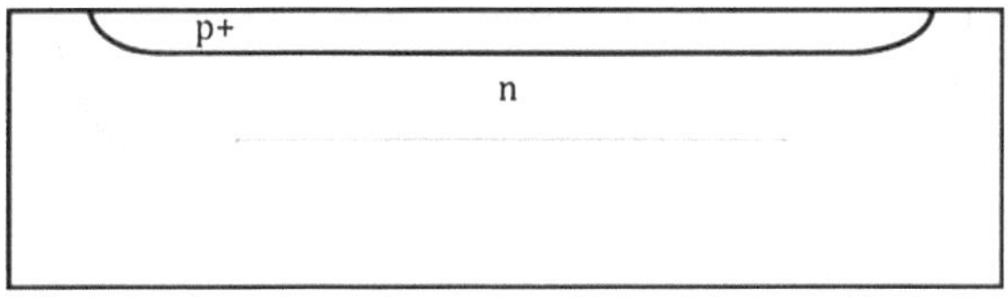

Fig. 2 Cross-sections of a generic pn junction in a planar process with the depletion region (*gray line*) forming in the structure upon reverse-biasing, assuming a large doping differential between the p and n regions

time distribution [5]. Once the initial carrier is in the depletion region, the statistics of impact ionization will create a time uncertainty that is roughly a normal distribution. The overall timing response of a SPAD to a pulsed laser can be described as a background noise plus a normal distribution convolved with the addition of an impulse function (describing carriers generated in the depletion region) with an exponential distribution.

3 CMOS SPAD Design

Building SPADs in CMOS substrates requires knowledge of the process and layers available to the designer to implement junctions that can be reverse-biased at high voltages. Figure 2 shows a generic pn junction implemented in a planar process. The figure shows the depletion region, as it forms upon reverse biasing the junction (assuming a large doping differential between the p and n regions).

Implementing a pn junction in a planar process first involves finding a way to prevent premature edge breakdown (PEB). Several techniques exist to implement PEB prevention. In essence, the techniques have in common the reduction of the electric field or the increase of the breakdown voltage at the edges of the junction, so as to maximize the probability that the avalanche is initiated in the center of the multiplication region, i.e. the region where the critical electric field for impact ionization is reached and, possibly, exceeded.

Figure 3 illustrates four of the most used structures. In (a) the n+ layer maximizes the electric field in the middle of the diode. In (b) the lightly doped p-implant reduces the electric field at the edge of the p+ implant. In (c) a floating p implant locally increases the breakdown voltage. A polysilicon gate is usually drawn to prevent the creation of a shallow trench, however, it can also be used to further extend the depletion region.

Shallow trench isolation (STI) can also be used to delimit the junction, provided that it is surrounded by a multi-layer of doped silicon so as to force recombination of those charges generated in the defect-rich STI as shown in structure (d) [6]. These structures are usually shaped as a ring around the junction; they are known as *guard rings*. Guard rings can also be defined *implicitly* by proper definition of drawn layers [7].

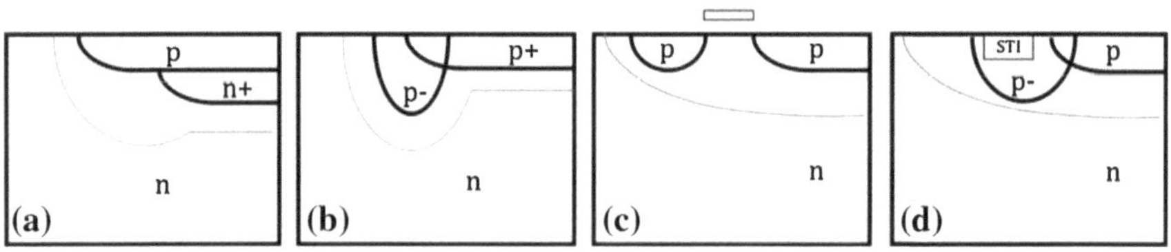

Fig. 3 Cross-sections of doping profiles that may be used to prevent premature edge breakdown in planar processes

There exist a variety of avalanche quenching techniques, partitioned in active and passive methods. The literature on these variants is extensive [8]. In active methods, the avalanche is detected and stopped by acting on the bias. In passive methods the pn junction bias is self-adjusted e.g. by a ballast resistor. Recharge methods can also be active and passive. In active methods, the bias across the diode is re-established by a switch activated by an avalanche detector. In passive methods the recharge occurs through the ballast.

Upon photon detection, the device generates a current pulse that is converted to a digital voltage level by means of a pulse shaping circuitry, also shown in the figure. The pulse shaper is also acting as an impedance adapter to drive the load of the column readout often employed in a SPAD matrix.

The main parameters characterizing individual SPADs are sensitivity, measured as **photon detection probability** (PDP), noise, measured as the rate of spurious pulses due to thermal events or **dark count rate** (DCR). Other parameters include **timing jitter**, also known somewhat inappropriately as **timing resolution**, **afterpulsing probability** and the aforementioned **dead time**. These parameters have been used in the literature for a variety of CMOS processes [9–17].

When implemented in an array, other performance measures become relevant to the quality of the imager. Dead time uniformity relates to the variability in dead time, which determines the dynamic range of each detector. Timing jitter uniformity and PDP uniformity, as well as DCR uniformity and crosstalk have to be accounted for and properly characterized [9]. PDP of course will also be a function of the input wavelength. In CMOS SPAD implementations, the sensitivity range is mostly in the visible spectrum, with a somewhat reduced near infrared and near ultraviolet PDP.

Crosstalk is also a chip-level effect similar to PDP and DCR non-uniformities; it relates to the interaction between an aggressor pixel and a victim pixel, where the aggressor may cause a spurious avalanche in a victim. The effect can be electrical and/or optical. Electrical crosstalk is due to electrical interference through substrate or supply noise. Optical crosstalk may occur when an avalanche is triggered in the aggressor; by impact ionization, several photons may be emitted, thus causing the victim to detect them. While electrical crosstalk is strongly dependent on the design of supply lines and of substrate noise rejection measures, optical crosstalk may only be influenced by the number of carriers involved in an avalanche and by pixel pitch.

4 TCSPC Based TOF Camera Systems

Using TCSPC for optical rangefinding in D-TOF mode has been proposed several decades ago, since the introduction of the LIDAR concepts. SPAD based single-pixel detectors, in combination with scanning, powerful pulsed light sources. One of the first examples of this combination was proposed in [18, 19]. In this work, the light source is synchronized to the SPAD to produce an accurate evaluation of the Time-Of-Flight of the reflected photons and thereby of the distance to the target.

In [20] this concept was made scannerless thanks to the use of monolithic arrays of SPADs implemented in CMOS technology. The concept, described in Fig. 4, is enabled by the use of a cone of light reaching approximately simultaneously the target source.

The photons reflected by the surface are imaged through a lens system to the array. With the use of an accurate chronometer or stop watch, it is possible to derive the distance of the reflecting point using the following relation

$$d \cong \frac{c}{2} \cdot \tau_{TOF}, \tag{2}$$

where τ_{TOF} is the Time-Of-Flight or the time difference between the light pulse synchronization signal and the time-of-arrival in the detector, and c is the speed of light in vacuum. Since SPADs are dynamic devices, they generate a digital pulse upon detection of a photon, and thus, unlike conventional diodes, they cannot hold a charge proportional to the overall photon count. Thus, the Time-Of-Flight must be computed in situ (either on pixel, on column, or on chip) or outside the image sensor. The same holds with integrated time-of-arrival evaluation: it can only be (1) in-pixel, (2) in-column, or (3) on-chip. To address this limitation, researchers have adopted a number of architectures that take advantage of the low propagation delay or high level of miniaturization achievable in standard submicron and deep-submicron CMOS technologies.

The simplest readout architecture implementing photon counting on-chip in combination with random-access single-photon detection, was demonstrated for the first time in a matrix of 32 × 32 pixels, each with an independent SPAD, a

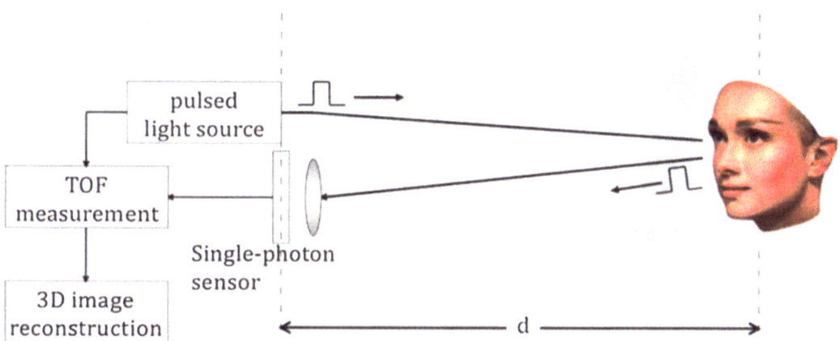

Fig. 4 Time-correlated single-photon counting (TCSPC) for optical rangefinding and 3D imaging

quenching mechanism, a pulse shaping and column access circuitry [20]. In this readout scheme, all time-sensitive operations had to be performed sequentially. This design has the main drawback in that it can only "see" one pixel at any point in time, while all the other pixels are operating but their output is lost. The micrograph of the chip is shown in Fig. 5; the chip was implemented in a 0.8 μm high-voltage CMOS process.

Addressing the readout bottleneck required some degree of sharing. The column is the obvious place to start from, since it is a repetitive structure that touches all pixel rows. The first known such approach involved interpreting the column as a bus. The bus is used as transfer time and address over the same hardware. The time is coded in partially processed pulses generated in the pixel of which the time-of-arrival is evaluated. The address is coded as a unique combination over the lines present in the bus sent to the bottom of the column where the time-of-arrival is evaluated, either off or on chip [11].

The second approach, known as latchless pipelined readout, has a passive type of coding, whereas time-of-arrival also contain the information of the row position where the pulse was generated. Every photon triggers a pulse that is injected onto the pipeline at a precise location that corresponds to the physical place where the pixel is located. Since the propagation time is across the column is too short to enable any time-based discrimination, a timing-preserving delay line is added to the column. At the bottom of the column time discrimination is performed by a column-based TDC that also returns the row code [21]. The photomicrograph

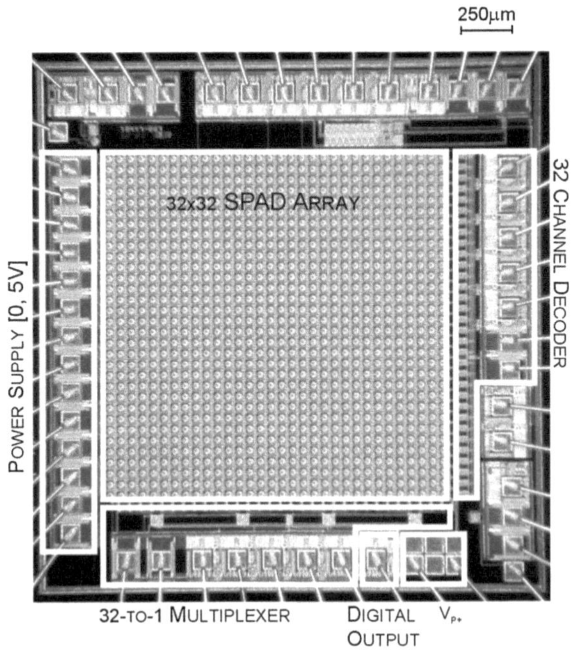

Fig. 5 The first large SPAD array with random access readout. The chip was implemented in 0.8 μm CMOS technology

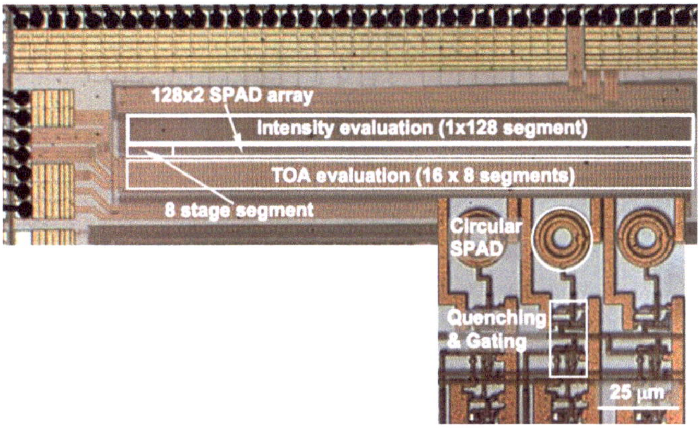

Fig. 6 CMOS version of a latchless pipeline based readout fabricated in 0.35 μm CMOS technology [21]

shown in Fig. 6 illustrates the actual implementation of the concept for a latchless pipelined 128 × 2 SPAD array fabricated in 0.35 μm CMOS.

An important step towards full parallelism was achieved with LASP [10], a 128 × 128 SPAD array, where a bank of 32 column-parallel time-to-digital converters (TDCs) was used to simultaneously process 128 in-line SPADs using an event-driven 4-to-1 column-multiplexer (one per column). Figure 7 shows the block diagram of LASP. Each TDC in the 32-array can generate 10 MS/s with a time resolution of 97 ps. The resolution can be further increased to 70 ps by acting on the clock frequency. Each TDC in LASP is a 3-tier partial TDC based on three different architectures: a clocked counter (2 MSBs, 25 ns resolution), a phase

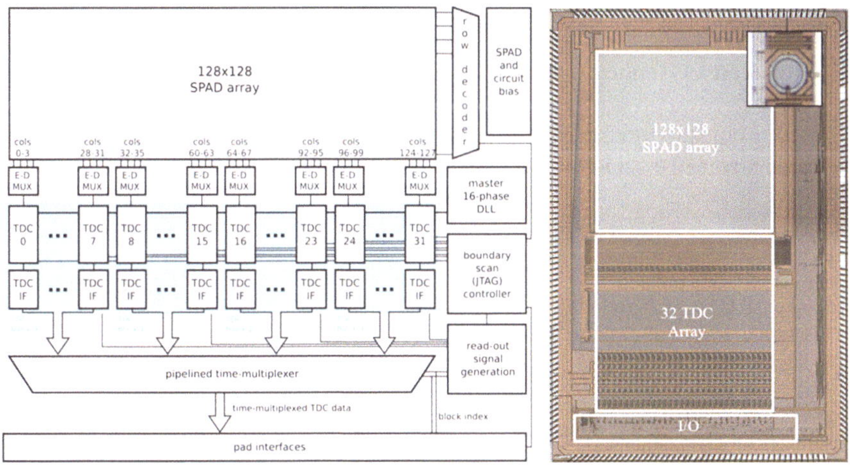

Fig. 7 Schematic and micrograph of LASP; an array of 32 TDCs processes the pulses generated by 128 SPADs at any time by means of a 4-to-1 event-driven multiplexer (one for each TDC)

interpolator controlled by a temperature-compensated DLL (4 intermediate bits, 1.56 ns resolution), and a 16-taps Vernier line (4 LSBs, 97.6 ps resolution). The total time resolution of 10bits is routed outside the chip through a high-speed digital network operating at 3.2 Gb/s. The differential non-linearity (DNL) and integral non-linearity (INL) were recently improved to ±0.1LSB and ±0.25LSB, respectively [22].

The chip was tested in TCSPC mode to compute a scene's depth via a pixel-by-pixel Time-Of-Flight evaluation of a defocused beam hitting the target. The results of the experiment are shown in the histogram of Fig. 8. The jitter is dominated by the SPAD timing uncertainty, whereas the characteristic tail of the device also appears from the picture.

The distance evaluation result is shown in Fig. 9 as a function of the real distance measured using a precision device from 40 cm to 3.75 m. Each distance measurement was derived from the centroid of the corresponding histogram, whereas the uncertainty as a function of distance is plotted in also plotted in the figure.

In Fig. 10 the resulting 3D image of a mannequin illuminated by a cone of pulsed laser light with a wavelength of 637 nm is shown after a 1 s exposure. The 3D points represent the centroid of the histograms of each pixel.

The integration of time-resolving electronics on a per-column or per-chip basis represents a trade-off between fill-factor, processing bandwidth and circuit area. Unlike integrating image sensors based on photodiodes, photons falling on SPAD pixels which are not multiplexed to available time-resolving channels are lost. In an efficient SPAD sensor operating in D-TOF mode, a time-stamp must be generated for every impinging photon at every pixel detector. This necessitates the combination of per-pixel time-digitization circuitry and high speed array readout.

The integration of a large array of in-pixel TDCs or time-to-amplitude converters (TACs) poses several challenges with respect to single channel architectures found in the literature [23–26]:

1. Circuit area is limited to a few 100 μm^2 to achieve practical pixel fill-factor and pitch.
2. Power consumption cannot exceed a few 100 μA per pixel, in order to allow array size to be scaled to 10's of kilopixels without excessive heating.

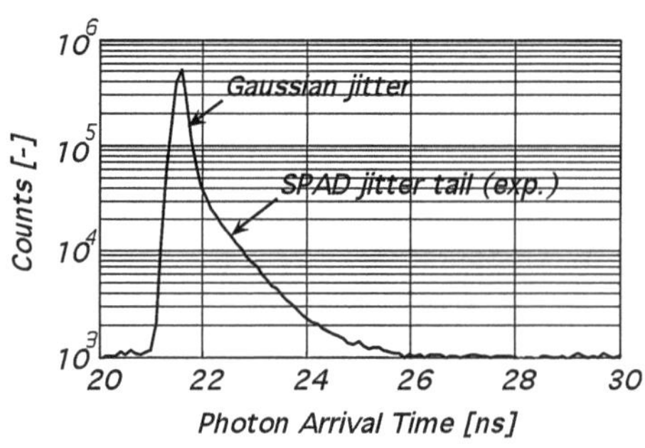

Fig. 8 TCSPC experiment. The laser pulse is pointed toward a SPAD in the array and the time-of-arrival of the first detected photon is evaluated by the corresponding TDC. The resulting histogram, shown in the plot in logarithmic scale, is then computed [10]

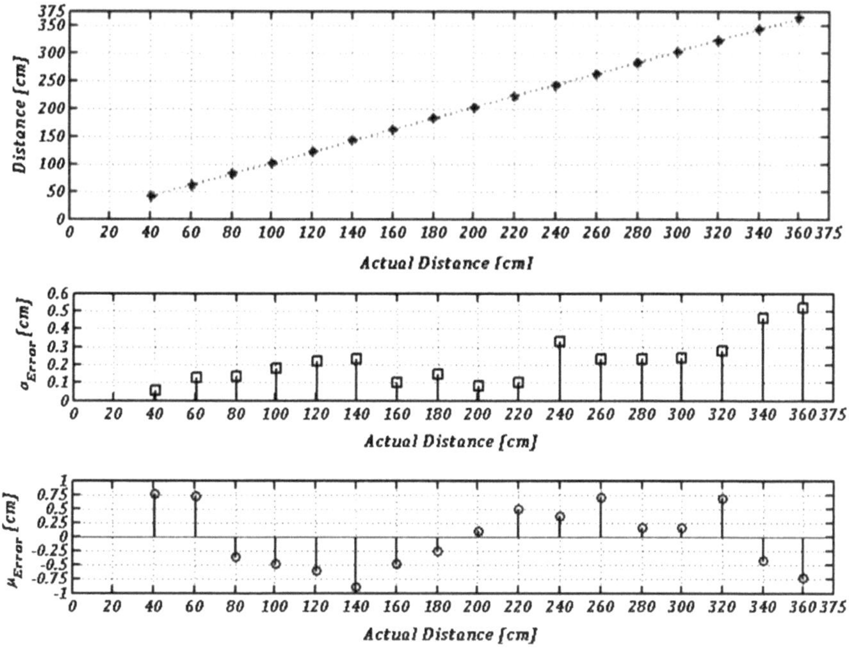

Fig. 9 Actual distance versus estimated distance computed in LASP at room temperature (*top*); average error versus distance (*middle*); standard deviation versus distance (*bottom*) [10]

Fig. 10 Target image computed by LASP in 1 s exposure at room temperature [10]

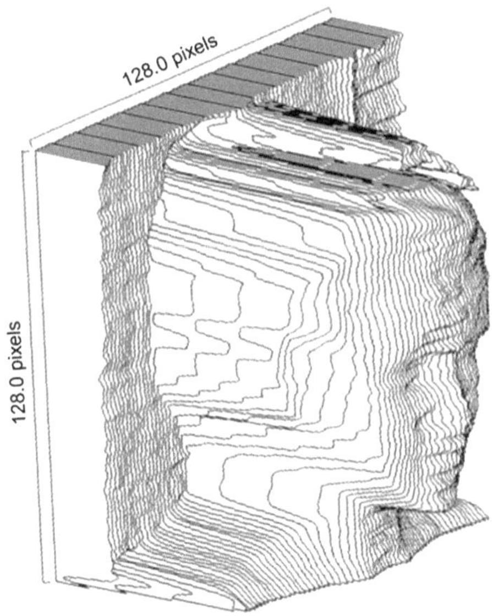

3. Uniformity requires around 1 percent matching of time resolution for acceptable image quality without requiring a frame store for pixel calibration.
4. Throughput must assure conversion of photon arrivals within typical pulsed illumination period of 100's of nanoseconds to avoid statistical distortions leading to non-linearity.
5. Time resolution is required to be a few 10's picoseconds in a full-scale range of around 25–50 ns resolution for indoor applications where human body features are to be distinguished at a maximum distance of a few meters.

These stringent specifications discount many of the conventional TDC architectures which achieve sub gate-delay resolution. Two scalable in-pixel TDC approaches were proposed in [27] and [28], differing by the adoption of an internal clock (IC-TDC) or an external clock (EC-TDC) as a source of timing accuracy (Fig. 11). Both converters operate in reverse START-STOP mode whereby they are started by a photon detected by the SPAD front-end circuit and stopped by a global clock synchronous with the pulsed optical source.

The IC-TDC integrates a gigahertz gated ring oscillator within each pixel, which clocks an n-bit ripple counter providing coarse photon arrival time estimates. Fine time estimates are obtained by decoding the frozen internal state of the ring providing the 3 least significant bits, which represent single inverter delays. A four-stage differential ring oscillator is chosen with a binary number of internal

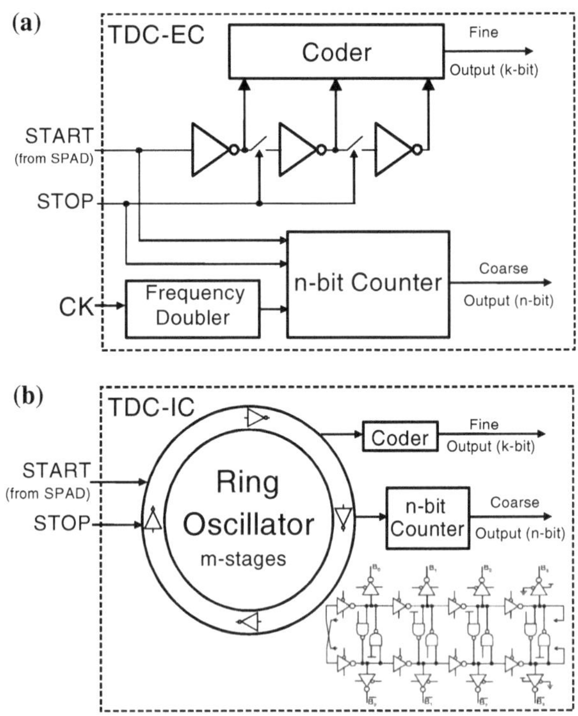

Fig. 11 Time-to-digital conversion pixels: **a** external clock **b** internal clock

states for simplicity of decoding logic. Two implementations have been studied, a chain of tri-stateable inverters provide static internal state memory [29] while the second, a chain of inverters connected by pass-gates provide a dynamic internal state memory [27]. Care must be taken to avoid metastability when gating the ring oscillator clock to the ripple counter, by employing hysteresis in the clock input stage. Tuning of the ring oscillator loop delay is achieved by current starving the differential inverters through a NMOS regulating transistor. This is also effective in diminishing the effect of supply noise induced jitter. IC-TDCs have the advantage of consuming power only when activated by an impinging photon, thus the power consumption of a large array is directly proportional to photon flux. This in turn has the potential disadvantage of introducing a light-flux-dependency of the TDC resolution, due to variable IR drops.

The EC-TDC achieves coarse time measurement in the similar manner to the IC-TDC, by means of an n-bit counter, clocked by an *external* high frequency clock distributed to the entire pixel array. Fine resolution is determined by the propagation delay of the SPAD pulse through an inverter chain. The state of inverter chain is frozen by the subsequent rising edge of the clock that is synchronized with the STOP signal and decode logic converts the thermometer code into a k-bit binary number. The STOP signal also halts the coarse counter. Time resolution calibration is performed by current-starving the inverter chain, through a feedback loop [28]. EC-TDCs consume an almost constant power level irrespective of input illumination but provide good global time resolution accuracy and negligible resolution dependency on photon flux.

Scaling of the per-pixel TDC approach from early prototypes of 32×32 to 160×128 pixels has been demonstrated with good performance. Characterization of a large IC-TDC image array has shown good DNL, INL, and pixel-to-pixel uniformity [29].

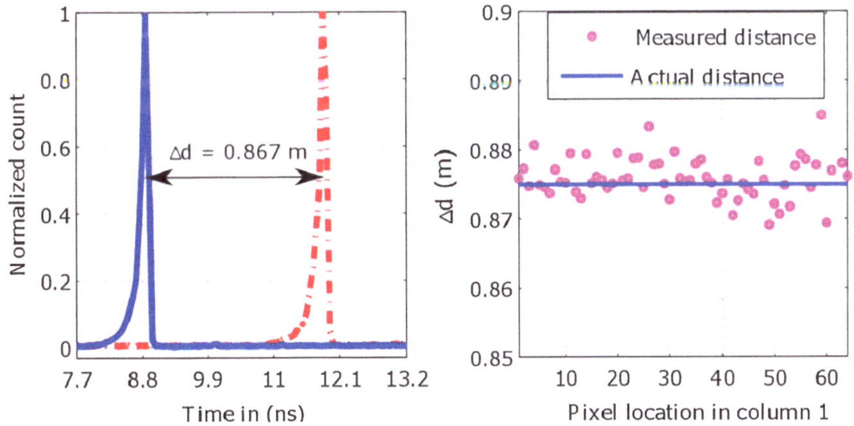

Fig. 12 Time of flight measurements with IC-TDC array

In Fig. 12, a TOF experiment is conducted for a single column of pixels, using a pulsed laser at a constant distance from the sensor. Future implementations of these architectures in advanced nanoscale CMOS technologies bring the prospect of considerable gains in fill factor, time resolution, power consumption, and pixel pitch.

Figure 13 shows an analog approach to per-pixel time estimation, employing an in-pixel TAC and ADC converter [30]. A current source (IbiasP) charges a capacitor Cs when the switch-structure, composed of Mp1, Mp2, Mp3, and IbiasN, is enabled. Three layout-matched replicas of the ramp-generator building block are employed: Stage1 and Stage2 are used alternately to measure the number of events or the event arrival time in a time interleaved way as Vo1 (or Vo2); StageREF is used to generate a reference voltage ramp Vo_{REF} for the embedded single-slope ADC. Analog time encoding is started on detection of a photon, by the charging of capacitor Cs, and stopped by the subsequent STOP clock rising edge. At the end of this interval, a voltage has accumulated on capacitor Cs that is proportional to the photon arrival time.

After an exposure time, the pixel array is switched over to analogue to digital conversion mode. The clock signal CNT and an n-bit Gray code count sequence operate synchronously and are globally distributed to the pixel array. The Stage-REF block generates a stepped analog ramp voltage Vo_{REF}, which is compared to

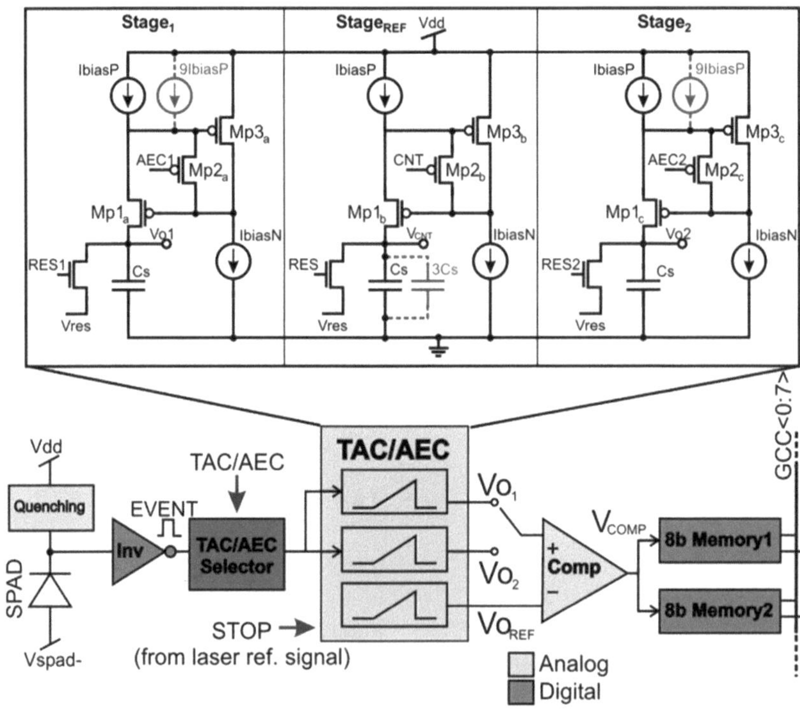

Fig. 13 Time-to-amplitude converter pixel architecture

Table 1 Performance comparison of per-pixel TAC/TDC arrays

Parameter	TAC [30]	TDC-EC [28]	TDC-IC [27]	Unit
Technology node	130	130	130	nm
Pixel pitch	50	50	50	μm
Bit resolution	6	10	10	bits
Time resolution (LSB)	160	119	178/52	ps
Uniformity	±2	±2	8	LSB
INL	1.9	±1.2	±0.4/1.4	LSB
DNL	0.7	±0.4	±0.5/2.4	LSB
Time jitter	<600	185	107/32	ps @ FWHM
Power consumption	300	94	28/38	μW @ 500kframe/sec

Vo1 (or Vo2) by an in-pixel comparator. When Vo_{REF} exceeds Vo1 (or Vo2) the voltage comparator toggles, sampling the Gray code into the memory. Hence the stored value of the Gray code is directly proportional to the analog voltage determined previously by the time-to-analog conversion process. Two memories are employed to allow one to participate in conversion while the other is being read-out. While the performance of the per-pixel TAC array was less favorable than the TDC counterparts (Table 1), this architecture shows greater promise in terms of scaling for area and power without the need for advanced nanoscale CMOS technology.

Highly parallel readout schemes must be employed to handle the extremely high volumes of per-pixel timestamp data generated by both per-pixel TDC and TAC image arrays. Single-photon detectors produce events at each pixel at much greater rates than conventional integrating image sensors, especially those generated by ambient lighting or DCR distribution. Data rates of several Gbps have been attained in recent imagers [29, 31].

Per-pixel time-resolving arrays, which integrated the TAC/TDC in same CMOS substrate as the SPAD detector face, have two practical drawbacks:

1. The bandgap of silicon limits detection to wavelengths below 1125 nm. Common CMOS SPADs have poor PDP (below 5 %) at suitable NIR wavelengths (850 or 930 nm) due to the shallow active junction.
2. The pixel fill factor is low due to the large insensitive area devoted to the time-resolving electronics; around 1–2 % have been demonstrated in the above research.

Hybrid 3D wafer technology offers a solution to this dilemma. Itzler et al. [32] have developed InGaAs/InP avalanche diode structures in the wavelength range of 0.92 to 1.67 μm, which are hybridized to CMOS read-out integrated circuits (ROICs) that enable independent laser radar Time-Of-Flight measurements for each pixel. The 9 % fill-factor of the SPAD array is improved to 75 % by employing a micro lens array and an average photon detection efficiency (PDE[3]) of

[3] PDE is defined as the multiplication of fill facto and PDP.

39 % is obtained. The ROIC consists of an array of 100 μm pitch pixels containing pseudorandom counters in a 0.18 μm CMOS process clocked from an external PLL distributed to all pixels. The power consumption of the array is only 50 mW at 20 kHz frame rate and 320 mW at 200 kHz. Aull et al. reported a similar 64 × 64, 50 μm pitch ROIC in 0.18 μm CMOS for a similar 3D stacked technology with a time resolution of 110 ps [33].

5 Single-Photon Synchronous Detection

The technique known as single-photon synchronous detection (SPSD) is an I-TOF measurement, the SPAD-digital equivalent of lock-in detection [34–38]. In SPSD, proposed for the first time in [39], the scene is illuminated with a sinusoidal or pseudo-sinusoidal light intensity and the phase of the return photons. As in lock-in systems, distance d is computed as

$$d \cong \frac{c\varphi}{2 \cdot 2\pi f_0}, \tag{3}$$

where c is the speed of light in vacuum, f_0 is the modulation frequency, and φ is the measured phase difference between outgoing signal and measured signal. In SPSD, unlike in lock-in systems, the phase φ is computed directly by counting photons in a structured way.

In SPSD the period of the illumination is discretized in N_C uniform time segments ΔT_0, $\Delta T_1, \ldots, \Delta T_k, \ldots, \Delta T_{N_{C-1}}$. At the end of the integration period, every counter stores a value C_k, corresponding to the counted photons in the time period ΔT_k during the integration period. For each segment the SPAD's output is directed to a counter that is activated only during that segment, the photon counter results C_k. shows how the illumination source is modulated and discretized. The figure also shows how the SPAD's output is distributed to the various counters to

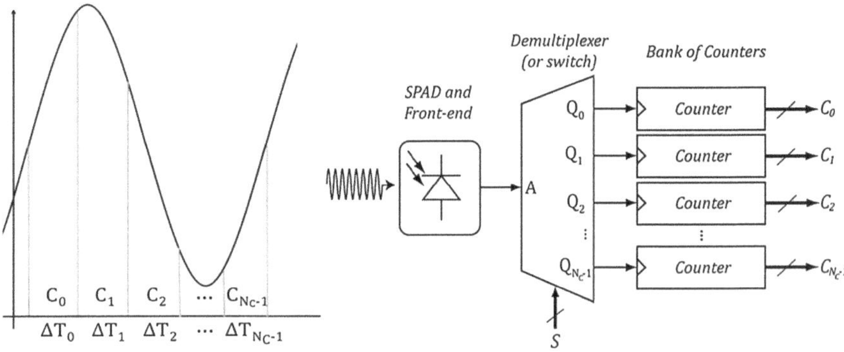

Fig. 14 Possible discretization of a period of illumination, with resulting partial counts (*left*); block diagram of the pixel in an SPSD camera as described in [39] (*right*)

generate C_0 through $C_{N_{C-1}}$ in a possible block diagram implementation of the technique on pixel (Fig. 14).

For the case $N_C = 4$, with only four partial counters C_0, C_1, C_2 and C_3, the phase becomes

$$\varphi = \arctan\left(\frac{C_3 - C_1}{C_0 - C_2}\right). \quad (4)$$

As a byproduct of this computation we also achieve an estimate of the offset and amplitude of the illumination intensity, as

$$A' = \frac{(C_3 - C_1)^2 + (C_0 - C_2)^2}{2}; \qquad B' = \frac{C_0 + C_1 + C_2 + C_3}{4}. \quad (5)$$

The parameters A′ and B′ are illustrated in Fig. 15 which shows a sequence of photon detection where the partial counts are progressively building up the waveform corresponding to a single period by superimposing adjacent measurements, thus enabling the computation of phase, offset, and amplitude.

A SPSD image sensor based on the concept outlined in Fig. 15 was implemented in an array of 60 × 48 pixels operating in real-time with an integration period as low as 10 ms and a frame rate as high as 100fps. A block diagram and a photomicrograph of the image sensor is shown in Fig. 16, whereas the pixel in this design comprises two 8-bit counters to compute alternatively C_0/C_2 and C_1/C_3.

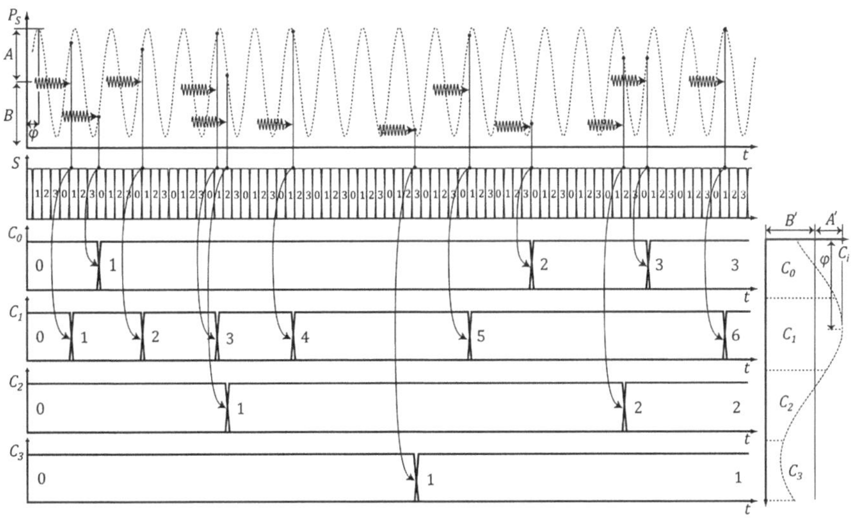

Fig. 15 Principle of operation of SPSD cameras. Impinging photons are detected and “assigned” to the appropriate bin depending upon the time-of-arrival with respect to the outgoing optical signal

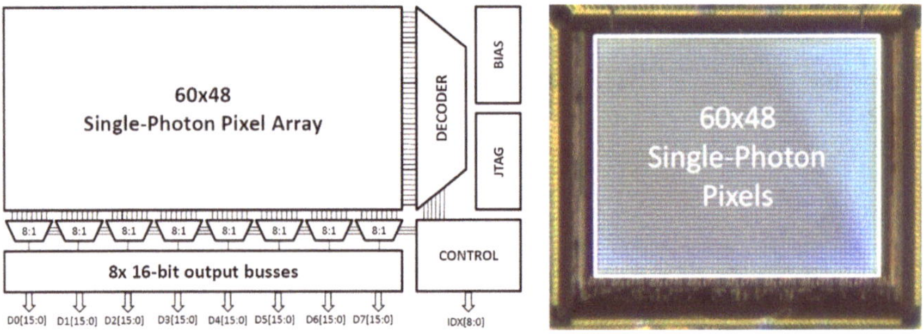

Fig. 16 Block diagram (*left*) and photomicrograph of the SPSD image sensor (*right*) proposed in [39] for the first time and fabricated in 0.35 μm CMOS technology

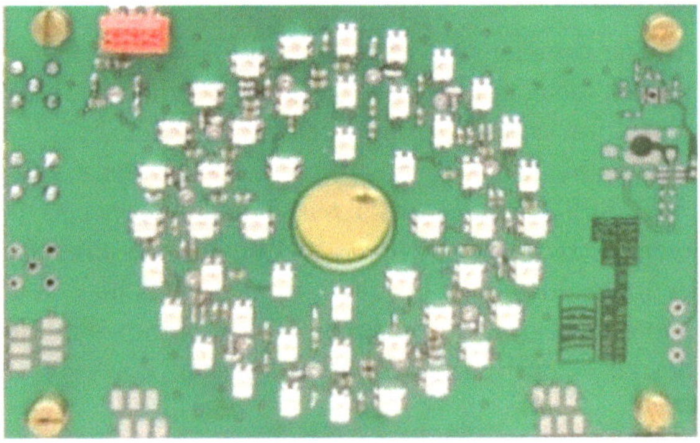

Fig. 17 Illuminator used in combination with the integrated SPSD image sensor. The total optical power emitted by the illuminator is 800 mW. The frequency of operation is 30 MHz, with a 3rd harmonic component suppressed at 37 dB

The pixel has a pitch of 85 μm and a fill factor of 1%, which can be compensated to a partial extent by the use of a microlens array [40].

The chip was used in several experiments to test its robustness to ambient light and interferences. The illuminator used was an array of 48 850 nm LEDs (Osram GmbH, Germany) located in three concentric circles as shown in Fig. 17.

The plots shown in Fig. 18 show the estimated distance versus the actual distance measured in a similar distance range as in the TCSPC experiments. The figure also shows mean errors and standard deviations as a function of the distance.

The same mannequin used in LASP was also used in the SPSD experiments. The results confirmed the estimated error. Figures 19 and 20 show a target at two exposure times, evidencing the relation between exposure, or number of the counting iterations, and overall distance error at the pixel level.

The images show the expected dependency of accuracy from exposure time. This behavior can be quantitatively analyzed looking at the demodulation contrast. Let us consider the statistical error of the distance measurement,

$$\sigma_{Error} = \frac{c}{2 \cdot 2\pi f_0} \sigma_{\varphi}, \tag{6}$$

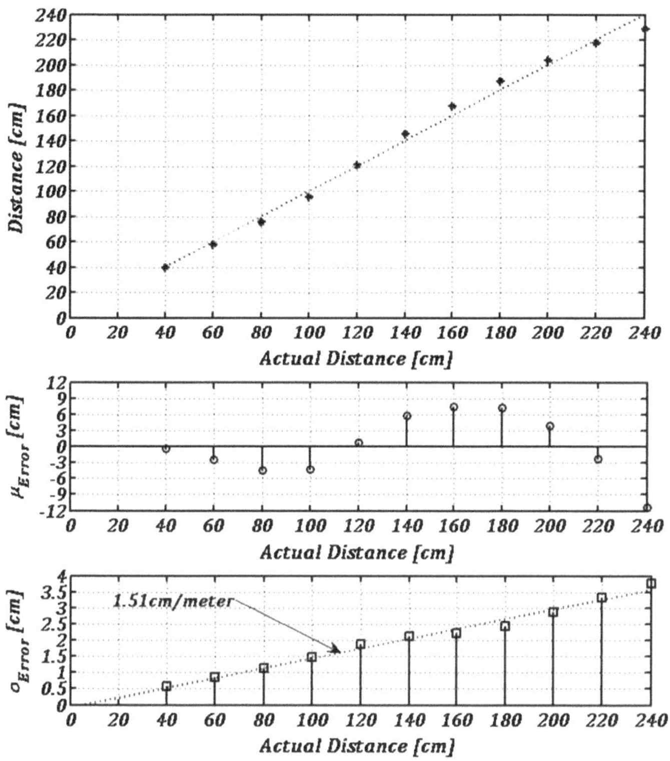

Fig. 18 Estimated versus actual distance using a SPSD pixel (*top*). Mean error versus distance (*middle*) and standard deviation of the error versus distance (*bottom*). All the data are reported at room temperature [39]

where σ_{φ} is the standard deviation of the phase measured using SPSD. Assuming $N_C = 4$, Eq. (6) can be rewritten in terms of the illumination parameters, as

$$\sigma_{Error} = \frac{R_D}{\sqrt{8}\pi}\frac{\sqrt{B'}}{A'}, \text{with}\, R_D = \frac{c}{2f_0}. \tag{7}$$

Rearranging the terms and substituting, we obtain the following equality

$$\sigma_{Error} \cong \frac{R_D}{\sqrt{8}\pi c_D}\frac{1}{\sqrt{T \cdot SBR \cdot S_R}}, \tag{8}$$

where T is the integration period, SBR is the signal-to-background ratio, and S_R is the signal count rate (the sum of all counters) on the image sensor. The term c_D is known as demodulation contrast, defined as

$$c_D = sinc\left(\pi\frac{\Delta T}{T_0}\right), \tag{9}$$

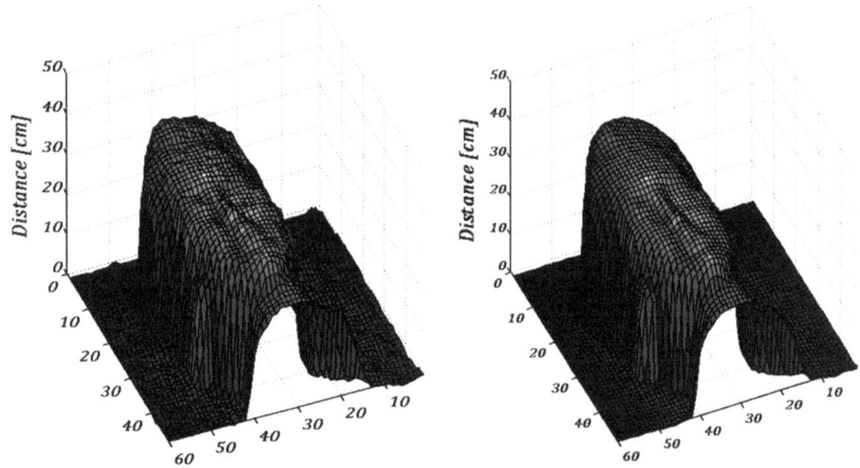

Fig. 19 3D image of mannequin in real-time exposure (30fps) at room temperature [39] at a different integration times

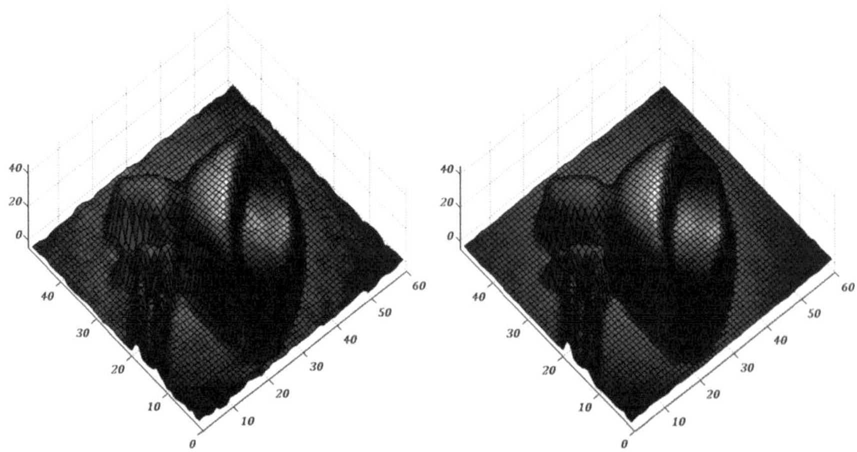

Fig. 20 Target image computed by SPSD in real-time exposure (30fps) at a shorter (50 ms) and longer (500 ms) integration times

where ΔT is the mean time segment duration, in this case ¼ of the total modulation period $T_0 = 1/f_0$. In this configuration, the demodulation contrast is 0.9, whereas 1.0 is the maximum achievable value. By increasing the number of samples to, say 8, the demodulation contrast would only increase to 0.974, but at a cost extra hardware and thus fill factor. Also note that the actual demodulation contrast decreases with modulation frequency. However, in SPSD it reduces very slowly while in lock-in systems it quickly degrades due to the limits in transit time in the pixels. The plot of Fig. 21 shows the relation between the demodulation contrast

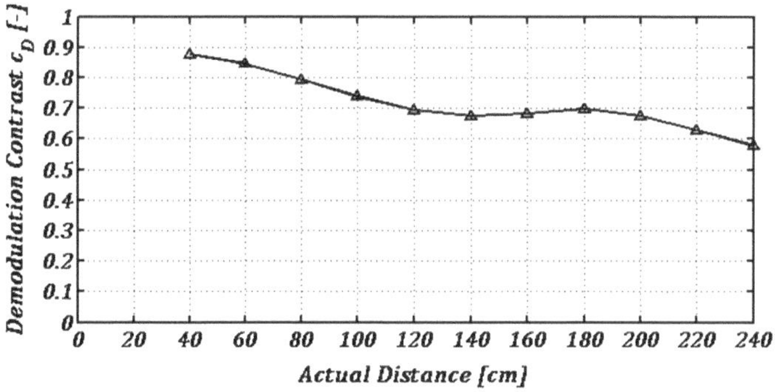

Fig. 21 Measured demodulation contrast as a function of distance in SPSD at room temperature [39]

and the availability of the signal S_R, as the target is further and the Time-Of-Flight increases.

6 Phase-Domain Delta-Sigma TOF Imaging

As has been discussed, phase-demodulating systems typically generate two or more bin values per frame requiring readout and external processing in order to produce a depth map. This IO and processing burden can result in a significant portion of the complete system's power consumption and computational load. Recent trends have seen an increasing level of complexity integrated into the sensor's focal plane to boost acquisition speed and reduce the required external circuitry. For example, while the first TOF 3D imaging systems were based on scanned single point range sensors comprising a discrete APD and external time-stamping circuitry, recent sensors have combined arrays of SPADs with on-chip time-stamping or binning logic, such as the TDC and SPSD systems discussed earlier in this chapter.

A natural extension of this theme of integration is the use of modern CMOS processes to construct sensors with additional focal-plane logic to allow the on-chip creation of depth-maps, reducing the IO and external processing needs of such systems. One such approach is to place a *Phase-Domain* Delta-Sigma (PD$\Delta\Sigma$) loop in the pixel. Figure 22 below compares the architecture of a simple $\Delta\Sigma$ ADC with its phase-domain equivalent. While the $\Delta\Sigma$ ADC measures the magnitude of an input *voltage* with respect to two reference voltages, the phase-domain implementation instead measures the mean *phase* of its input with respect to a pair of reference phases. This approach has been successfully demonstrated in applications including magnetic flux [41] and temperature [42] sensing, where the particular sensing elements used created a phase shift in response to the property to be measured.

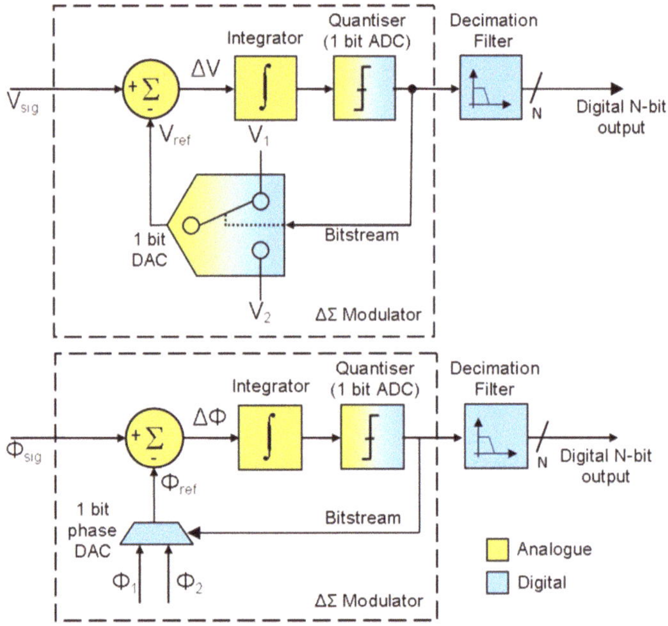

Fig. 22 Comparison of: conventional ΔΣ ADC (*top*), phase-domain ΔΣ converter for 3D imaging (*bottom*)

I-TOF 3D imaging systems also of course depend on the measurement of a phase shift. Crucially, an all-digital PDΔΣ loop may be constructed which locally processes the SPAD pulses created within a pixel and converges to estimate this depth related phase shift directly. This approach was employed in the 128 × 96 pixel sensor reported in [31], as illustrated in the architecture block diagram and die micrograph shown in Figs. 23 and 24 below.

The loop operates in a similar fashion to existing two-bin demodulating approaches [43], dividing the returning pulse energy into two bins: in phase and 180° out of phase with the transmitted modulated light. However, instead of simply integrating the energy in these bins over a period of time and transmitting their contents for external processing, the loop continually seeks to drive its internal integrator towards zero by integrating positively during φ_1 and negatively during φ_2, as the timing diagram contained in Fig. 25 below illustrates. The resulting bitstream density, r, indicates the mean phase of the input pulses, as governed by Eq. 10, where $N_{\varphi 1}$ and $N_{\varphi 2}$ represent the number of photons counted in bins φ_1 and φ_2 respectively.

$$r = \frac{N_{\phi_2}}{N_{\phi_1} + N_{\phi_2}} \tag{10}$$

This calculation is of course influenced by the DC component of the detected light, which must be corrected, for example by subtracting the DC component

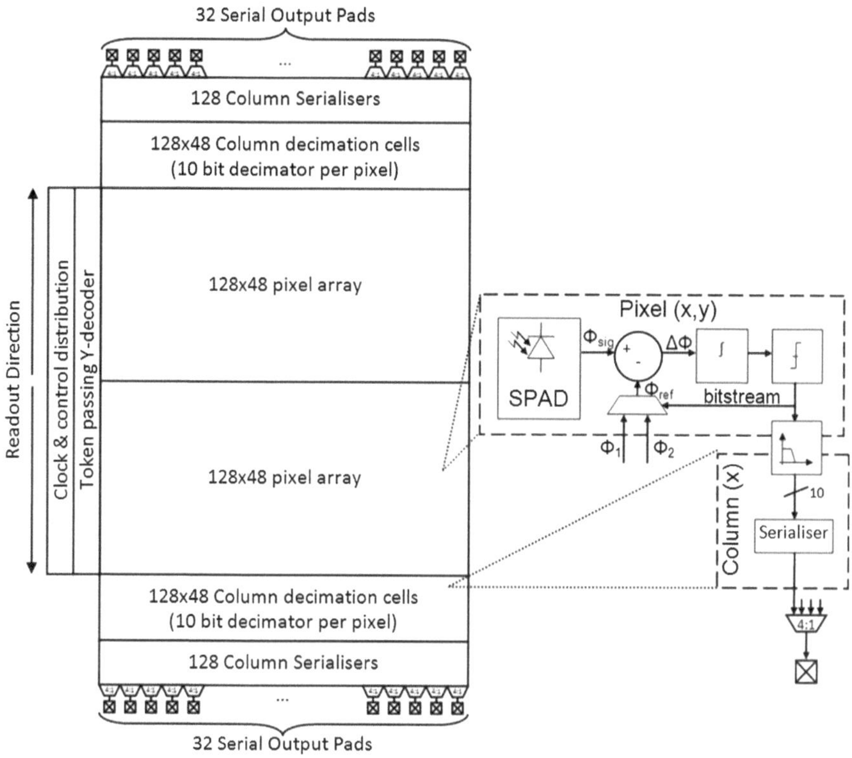

Fig. 23 Architecture of 128 × 96 pixel, all-digital, phase-domain ΔΣ based sensor reported in [31]

from each bin value using an integration of equal exposure time with the illumination source disabled.

Characterization data for the PDΔΣ, including INL and repeatability error characterization was presented in [31]. Relatively low illumination power and pulse rates were used at the expense of repeatability error. However, the PDΔΣ approach achieved an excellent linearity of ±5 mm over a 0.4–2.4 m range, with photon pile up being the dominant effect limiting accuracy at very short distances.

Figure 26 shows example images captured using the PDΔΣ sensor, serving as the first demonstration of the on-chip creation of depth-maps with only simple inter-frame averaging and defect correction being applied externally before rendering.

The use of an in-pixel, phase-domain ΔΣ loop to directly sense the mean phase of incident photons detected synchronously to the modulation of the outgoing illumination light is one approach to the management of the large volumes of data produced by single-photon 3D imagers and the associated computational burden on the host system. With the increasing commercial relevance of 3D imaging systems, such techniques seem likely to continue to develop, facilitated by the

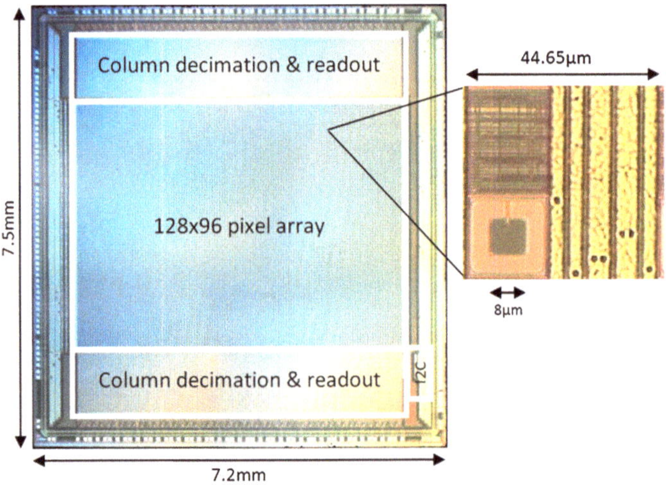

Fig. 24 Die micrograph of 128 × 96 pixel, all-digital, phase-domain ΔΣ based sensor reported in [31]

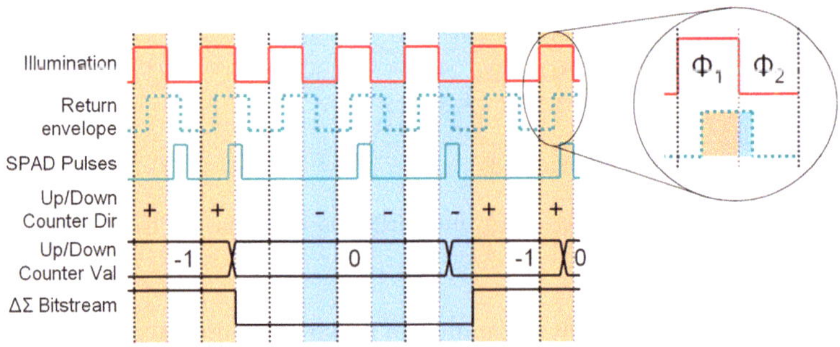

Fig. 25 Timing diagram of all-digital, phase-domain ΔΣ based pixel [31]

capabilities of modern CMOS imaging processes to yield highly integrated imaging systems.

7 Perspectives and Outlook

More recently, scanned 3D cameras have appeared based on relatively small arrays of relatively large pixels whose core is a so-called silicon photomultiplier (SiPM). SiPMs are essentially an array of non-imaged SPADs connected so as to sum all avalanche currents in one point. The more recent digital SiPM replaces the analog summing node with an OR operation. The advantage of d-SiPMs is a faster response to photon bunches and the capability of turning off individual noisy

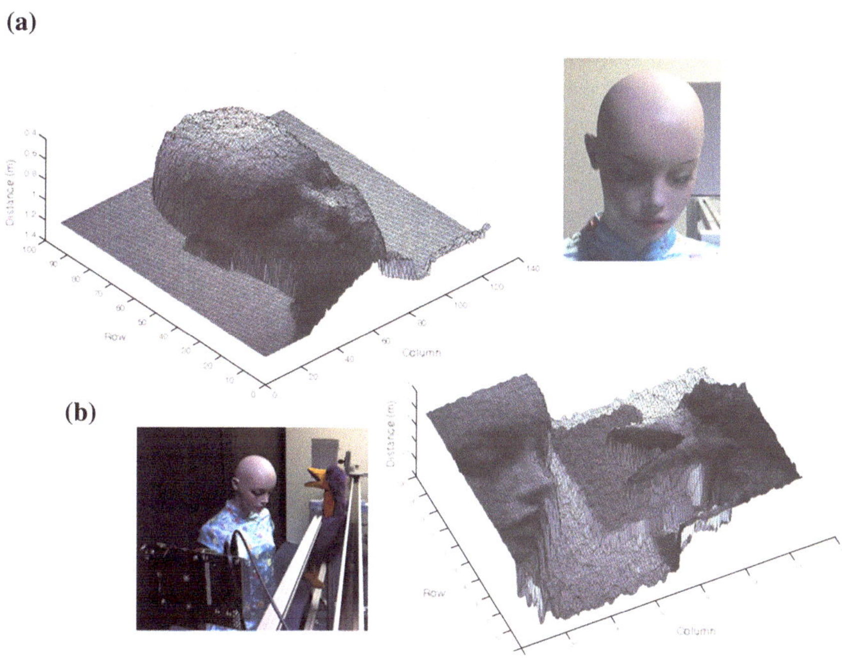

Fig. 26 Example images captured using PDΔΣ sensor [31]: **a** 20 second exposure showing mannequin at 1m distance. **b** 1 second exposure showing objects at 0.75m, 1.1m, and 1.8m

SPADs. An example of this trend has been reported by [44], where four 10×10 mini SiPMs have been implemented in 0.18 μm HV CMOS technology in combination with a mirror to scan large areas in TCSPC mode.

The use of mini SiPMs is spreading to other fields and applications. An example is that of medical sensors for positron emission tomography that rely on time-of-arrival detection. Borrowing technologies developed in 3D camera technology, more TDC integration together with SPADs and SPAD arrays is a clear trend.

Recent SPAD structures demonstrated in nanoscale CMOS exhibit improving DCR and spectral efficiency, as well as compatibility with TSV and backside processing [45–47]. Adoption of these advanced processes brings the prospect of simultaneous improvements in time resolution, fill factor, pixel pitch as well as the capacity to integrate on-chip Time-Of-Flight computation to ease I/O data rate demands. Analog approaches to time-resolved SPAD pixels offer a route to the smallest pixel pitch provided uniformity issues are addressed [48].

On another front, the emergence of III-V materials in configurations that are fully compatible with a CMOS fabrication line may bring these materials to the mainstream in 3D imaging. Examples of this trend are two independent works reporting the first Ge-on-Si SPADs fabricated in a way that is fully compatible with a conventional CMOS technology [49, 50]. More activity in this domain is expected, especially in the creation of larger and more compact arrays.

Finally, 3D integration of CMOS devices is becoming a commercial reality, hence, we expect that it will extend to 3D image sensors based on SPADs where through silicon via and backside illuminated devices will become routine in the near future. This will have an immediate impact on fill factor and imager sizes. Later cost will be positively impacted and packaging simplified.

Acknowledgments The authors are grateful to the Swiss National Science Foundation, the Swiss Government sponsored Nano-Tera and MICS grants, and Xilinx Inc.'s University Program.

References

1. S. Cova, A. Longoni, A. Andreoni, Towards picosecond resolution with single-photon avalanche diodes. Rev. Sci. Instr. **52**(3), 408–412 (1981)
2. R.J. McIntyre, Recent developments in silicon avalanche photodiodes. Measurement **3**(4), 146–152 (1985)
3. S.M. Sze, Physics of Semiconductor Devices, 2nd edn. (Wiley-Interscience, New York, 1981)
4. A. Spinelli, A. Lacaita, Physics and numerical simulation of single photon avalanche diodes. IEEE Trans. Electron Devices **44**, 1931–1943 (1997)
5. G. Ripamonti, S. Cova, Carrier diffusion effects in the time-response of a fast photodiode. Solid State Electron. **28**(9), 925–931 (1985)
6. M. Gersbach, J. Richardson, E. Mazaleyrat, S. Hardillier, C. Niclass, R.K. Henderson, L. Grant, E. Charbon, A low-noise single-photon detector implemented in a 130 nm CMOS imaging process. Solid-State Electron. **53**(7), 803–808 (2009)
7. J. Richardson, L. Grant, R. Henderson, A low-dark count single-photon avalanche diode structure compatible with standard nanometer scale CMOS technology, *International Image Sensor Workshop* (2009)
8. S. Cova, M. Ghioni, A. Lacaita, C. Samori, F. Zappa, Avalanche photodiodes and quenching circuits for single-photon detection. Appl. Opt. **35**(12), 1956–1976 (1996)
9. A. Rochas et al., Single photon detector fabricated in a complementary metal–oxide–semiconductor high-voltage technology. Rev. Sci. Instr. **74**(7), 3263–3270 (2003)
10. C. Niclass, C. Favi, T. Kluter, M. Gersbach, E. Charbon, A 128x128 Single-Photon Image Sensor with Column-Level 10-bit Time-to-Digital Converter Array. IEEE J. Solid-State Circuits **43**(12), 2977–2989 (2008)
11. C. Niclass, M. Sergio, and E. Charbon, A single photon avalanche diode array fabricated in deep-submicron CMOS technology, IEEE Design, Automation and Test in Europe, 1–6 (2006)
12. H. Finkelstein, M.J. Hsu, S.C. Esener, STI-Bounded single-photon avalanche diode in a deep-submicrometer CMOS technology. IEEE Electron Device Lett. **27**(11), 887–889 (2006)
13. C. Niclass, M. Sergio, E. Charbon, *A Single Photon Avalanche Diode Array Fabricated in 0.35μm CMOS and based on an Event-Driven Readout for TCSPC Experiments* (SPIE Optics East, Boston, 2006)
14. D. Stoppa, L. Pacheri, M. Scandiuzzo, L. Gonzo, G.-F. Della Betta, A. Simoni, A CMOS 3-D imager based on single photon avalanche diode. IEEE Trans. Circuits Syst. **54**(1), 4–12 (2007)
15. L. Pancheri, D. Stoppa, Low-noise CMOS Single-photon Avalanche Diodes with 32 ns Dead Time, in *IEEE European Solid-State Device Conference* (2007)

16. N. Faramarzpour, M.J. Deen, S. Shirani, Q. Fang, Fully integrated single photon avalanche diode detector in standard CMOS 0.18-um technology. IEEE Trans. Electron Devices **55**(3), 760–767 (2008)
17. C. Niclass, M. Sergio, E. Charbon, A CMOS 64x48 single photon avalanche diode array with event-driven readout, in *IEEE European Solid-State Circuit Conference* (2006)
18. J.S. Massa, G.S. Buller, A.C. Walker, S. Cova, M. Umasuthan, A.M. Wallace, Time-pf-flight optical ranging system based on time-correlated single-photon counting. App. Opt. **37**(31), 7298–7304 (1998)
19. M.A. Albota et al., Three-dimensional imaging laser radars with Geiger-mode avalanche photodiode arrays. Lincoln Labs J. **13**(2) (2002)
20. C. Niclass, A. Rochas, P.A. Besse, E. Charbon, Design and characterization of a CMOS 3-D image sensor based on single photon avalanche diodes. IEEE J. Solid-State Circuits **40**(9), 1847–1854 (2005)
21. M. Sergio, C. Niclass, E. Charbon, A 128x2 CMOS single photon streak camera with timing-preserving latchless pipeline readout, in *IEEE International Solid-State Circuits Conference* (2007), pp. 120–121
22. J.M. Pavia, C. Niclass, C. Favi, M. Wolf, E. Charbon, 3D near-infrared imaging based on a SPAD image sensor, *International Image Sensor Workshop* (2011)
23. J.-P. Jansson et al., A CMOS time-to-digital converter with better than 10 ps single-shot precision. IEEE J. Solid-State Circuits **41**(6), 1286–1296 (2006)
24. A.S. Yousif et al., A fine resolution TDC architecture for next generation PET imaging. IEEE Trans. Nucl. Sci. **54**(5), 1574–1582 (2007)
25. M. Lee, et al., A 9b, 1.25 ps resolution coarse-fine time-to-digital converter in 90 nm cmos that amplifies a time residue, in *IEEE Symposium on VLSI Circuits* (2007), pp. 168–169
26. P. Chen et al., A low-cost low-power CMOS time-to-digital converter based on pulse stretching. IEEE Trans. Nucl. Sci. **53**(4), 2215–2220 (2006)
27. J. Richardson, R. Walker, L. Grant, D. Stoppa, F. Borghetti, E. Charbon, M. Gersbach, R.K. Henderson, A 32x32 50 ps resolution 10 bit time to digital converter array in 130 nm CMOS for time correlated imaging, in *IEEE Custom Integrated Circuits Conference* (2009), pp. 77–80
28. M. Gersbach, Y. Maruyama, E. Labonne, J. Richardson, R. Walker, L. Grant, R.K. Henderson, F. Borghetti, D. Stoppa, E. Charbon, A Parallel 32x32 time-to-digital converter array fabricated in a 130 nm imaging CMOS technology, in *IEEE European Solid-State Device Conference* (2009)
29. C. Veerappan, J. Richardson, R. Walker, D.U. Li, M.W. Fishburn, Y. Maruyama, D. Stoppa, F. Borghetti, M. Gersbach, R.K. Henderson, E. Charbon, A 160x128 single-photon image sensor with on-pixel, 55 ps 10b time-to-digital converter,in *IEEE International Solid-State Circuits Conference* (2011), pp. 312–314
30. D. Stoppa, F. Borghetti, J. Richardson, R. Walker, L. Grant, R.K. Henderson, M. Gersbach, E. Charbon, A 32x32-pixel array with in-pixel photon counting and arrival time measurement in the analog domain, in *IEEE European Solid-State Device Conference* (2009), pp. 204–207
31. R.J. Walker, J.R. Richardson, R.K. Henderson; A 128 $\times$ 96 pixel event-driven phase-domain $\Delta\Sigma$-Based fully digital 3D camera in 0.13 μm CMOS imaging technology", in *IEEE International Solid-State Circuits Conference* (2011), pp. 410–412
32. M.A. Itzler, M. Entwistle, M. Owens, K. Patel, X. Jiang, K. Slomkowski, S. Rangwala, Geiger-mode avalanche photodiode focal plane arrays for three-dimensional imaging LADAR, in *SPIE Infrared Remote Sensing and Instrumentation* (2010), p. 7808
33. B. Aull, J. Burns, C. Chenson, B. Felton, H. Hanson, C. Keast, J. Knecht, A. Loomis, M. Renzi, A. Soares, S. Vyshnavi, K. Warner, D. Wolfson, D. Yost, D. Young, Laser radar imager based on 3D integration of Geiger-Mode avalanche photodiodes with two SOI timing circuit layers, in *Proceedings of the IEEE International Solid-State Circuits* (2006), 304–305
34. T. Spirig, P. Seitz, O. Vietze, F. Heitger, The lock-in CCD-two-dimensional synchronous detection of light. IEEE J. Quantum Electron. **31**(9), 1705–1708 (1995)

35. R. Miyagawa, T. Kanade, CCD-based range-finding sensor. IEEE Trans. Electron Devices **41**(10), 1648–1652 (1997)
36. R. Lange, P. Seitz, Solid-state Time-Of-Flight range camera. IEEE J. Quantum Electron. **37**(3), 390–397 (2001)
37. R. Schwarte, Z. Xu, H. Heinol, J. Olk, B. Buxbaum, New optical four-quadrant phase detector integrated into a photogate array for small and precise 3D cameras. SPIE Three-Dimensional Image Capture **3023**, 119–128 (1997)
38. C. Bamji, E. Charbon, Systems for CMOS-compatible three-dimensional image sensing using quantum efficiency modulation, US Patent 6,580,496 (2003)
39. C. Niclass, C. Favi, T. Kluter, F. Monnier, E. Charbon, Single-photon synchronous detection. IEEE J. Solid-State Circuits **44**(7), 1977–1989 (2009)
40. S. Donati, G. Martini, M. Norgia, Microconcentrators to recover fill-factor in image photodetectors with pixel on-board processing circuits. Opt. Express **15**(26), 18066–18075 (2007)
41. S. Kawahito, A. Yamasawa, S. Koga, Y. Tadokoro, K. Mizuno, O. Tabata; Digital interface circuits using sigma-delta modulation for integrated micro-fluxgate magnetic sensors, in *IEEE International Symposium on Circuits and Systems*, vol. 4 (1996), pp. 340–343
42. C. van Vroonhoven, K. Makinwa, A CMOS temperature-to-digital converter with an inaccuracy of $\pm$ 0.5°C (3σ) from -55 to 125°C, in *IEEE International Solid-State Circuits Conference* (2008), pp. 576–637
43. S. Kawahito, I.A. Halin, T. Ushinaga, T. Sawada, M. Homma, Y. Maeda, A CMOS Time-Of-Flight range image sensor with gates-on-field-oxide structure. IEEE Sens. J. **7**(12), 1578–1586 (2007)
44. C. Niclass, M. Soga, S. Kato, A 0.18 μm CMOS single-photon sensor for coaxial laser rangefinders, in *Asian Solid-State Circuits Conference* (2010)
45. M. Karami, M. Gersbach, E. Charbon, A new single-photon avalanche diode in 90 nm standard CMOS technology, SPIE Optics + Photonics, NanoScience Engineering, Single-Photon Imaging (2010)
46. R.K. Henderson, E. Webster, R. Walker, J.A. Richardson, L.A. Grant, A 3x3, 5um Pitch, 3-transistor single photon avalanche diode array with integrated 11 V bias generation in 90 nm CMOS technology, in *IEEE International Electron Device Meeting* (2010), pp. 1421–1424
47. E.A.G.Webster, J.A. Richardson, L.A. Grant, D. Renshaw, R.K. Henderson, An infra-red sensitive, low noise, single-photon avalanche diode in 90 nm CMOS. International Image Sensor Workshop (IISW), Hokkaido, 8–11 June 2011
48. L. Pancheri, N. Massari, F. Borghetti, D. Stoppa, A 32x32 SPAD pixel array with nanosecond gating and analog readout. International Image Sensor Workshop (IISW), Hokkaido, 8–11 June 2011
49. A. Sammak, M. Aminian, L. Qi, W.D. de Boer, E. Charbon, L. Nanver, A CMOS Compatible Ge-on-Si APD Operating in Proportional and Geiger Modes at Infrared Wavelengths, in *International Electron Device Meeting* (2011)
50. Z. Lu, Y. Kang, C. Hu, Q. Zhou, H.-D. Liu, J.C. Campbell, Geiger-mode operation of Ge-on-Si avalanche photodiodes. IEEE J. Quantum Electron. **47**(5), 731–735 (2011)

Electronics-Based 3D Sensors

Matteo Perenzoni, Plamen Kostov, Milos Davidovic, Gerald Zach and Horst Zimmermann

The conventional photodiode, available in every CMOS process as a PN junction, can be enriched by smart electronics and therefore achieve interesting performance in the implementation of 3D Time-Of-Flight imagers. The high level of integration of deep submicron technologies allows the realization of 3D pixels with interesting features while keeping reasonable fill-factors.

The pixel architectures described in this chapter are "circuit-centric", in contrast with the "device-centric" pixels that will be introduced in the chapter talking about photo-mixing devices chap. 4. The operations needed to extract the 3D measurement are performed using amplifiers, switches, and capacitors. The challenges to face with this approach are the implementation of per-pixel complex electronic circuits in a power- and area-efficient way.

Several approaches have been pursued, based on modulated or pulsed light, with pros and cons: the modulated light allows exploiting the high linearity of the technique while the pulsed light enables fast 3D imaging and increased signal-to-noise ratio. Different techniques will be introduced aimed at improving the rejection to background light, which is a strong point of the electronics-based 3D pixels.

M. Perenzoni (✉)
Smart Optical Sensors and Interfaces, Fondazione Bruno Kessler,
via Sommarive 18 I-38100 Povo (TN), Italy
e-mail: perenzoni@fbk.eu

P. Kostov · M. Davidovic · G. Zach · H. Zimmermann
Institute of Electrodynamics, Microwave and Circuit Engineering,
Vienna University of Technology, Gußhausstr. 25/354 1040 Vienna, Austria

M. Davidovic
Avago Technologies Fiber Austria GmbH, Webergasse 18 1200 Vienna, Austria

G. Zach
RIEGL Laser Measurement Systems GmbH, 3580 Horn, Austria

F. Remondino and D. Stoppa (eds.), *TOF Range-Imaging Cameras*,
DOI: 10.1007/978-3-642-27523-4_3,

1 Pulsed I-TOF 3D Imaging Method

In the following a detailed theoretical analysis of 3D imaging based on pulsed indirect Time-Of-Flight (I-TOF) will introduce the operating principle and the main relationships between the physical quantities. The first pioneering work on electronics-based sensors for 3D is described in [1] and employed the pulsed indirect TOF technique explained in the following paragraphs.

1.1 Principle of Operation

Pulsed I-TOF 3D imagers rely on the indirect measurement of the delay of a reflected laser pulse to extract the distance information. Several integration windows can be defined, leading to a number of integrated portions of the reflected laser pulse: each window, properly delayed, can map a different distance range [2].

In particular, as depicted in Fig. 1, the principle of operation exploits the integration of a "slice" of the reflected pulse during a temporal window (W_1), whose beginning determines the spatial offset of the measurement. So, the more the laser pulse is delayed (the farther the object), the larger will be the integrated value. Since the pulse amplitude depends also on the object reflectivity and distance, the same measure must be done also with a window that allows complete integration of the laser pulse (W_0), for either near or far objects. Then the latter value can be used to normalize the integral of the sliced pulse and to obtain a value that is linearly correlated to the distance. Note that in W_1 also the complementary part of the pulse can be used, with similar results [3].

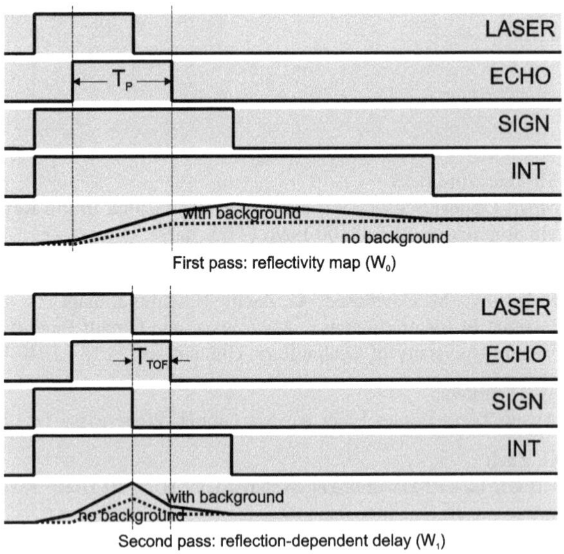

Fig. 1 Description of the pulsed Time-Of-Flight waveforms

The removal of background light can be performed by changing the polarity of the integration during the windows W_0 and W_1: the background contribution is cancelled at the end of the integration.

The realization of a 3D frame is obtained by accumulation of several laser pulses for each window in order to increase the signal-to-noise ratio of the integrated voltage: for a given frame rate this number of accumulations is typically limited by the specifications of the illuminator, which usually are lasers with a maximum duty-cycle of about 0.1 %.

Ideally, if the laser pulse, of length T_p, is fired with no delays with respect to the beginning of window W_0, the following equation allows the distance measurement to be extracted:

$$z_{TOF} = \frac{c \cdot T_p}{2} \cdot \frac{V_{W1}}{V_{W0}} = z_{\max} \cdot \frac{V_{W1}}{V_{W0}} \tag{1}$$

The maximum measurable distance z_{max} with two integration windows is defined by the length of the pulse and requires a minimum delay between W_0 and W_1 of T_p, while the length of the integration windows is at minimum equal to $2T_p$ if the background subtraction is implemented.

1.2 Signal and Noise Analysis

The performance of an electronics-based pulsed I-TOF system can be theoretically predicted; the considerations start from the power budget and optics: given a power density P_d on a surface, the power hitting the pixel is obtained as follows:

$$P_{pix} = \frac{\tau_{opt} A_{pix} FF}{4F\#^2} P_d \tag{2}$$

where FF is the pixel fill-factor, A_{pix} is the area, τ_{opt} is the optics transmission and F# is the optics f-number.

While the background power density depends on the illumination level, the reflected laser pulse amplitude depends on the reflectivity ρ, distance z and beam divergence θ; so, with the hypothesis that the laser beam projects the total power P_{laser} on a square, the power density on the target can be expressed as:

$$P_{dlaser} = \frac{\rho \cdot P_{laser}}{(2z \tan \theta/2)^2} \tag{3}$$

Background and laser pulses contribute each other to the generated electrons:

$$N(T_{\text{int}}) = \frac{QE(\lambda) \cdot T_{\text{int}}}{hc/\lambda} \tag{4}$$

where the wavelength λ, quantum efficiency QE and integration time T_{int} may be different for the two contributions. As already shown in (1), since the output voltage is obtained integrating on the same capacitance, the distance can be calculated from the charge integrated during W_0 and W_1:

$$z_{TOF} = z_{\max} \cdot \frac{Q_{W2}}{Q_{W1}} \tag{5}$$

.

The background contributes ideally with zero charge to the total amplitude, so for m accumulations the number of electrons N_{laser} from eq. (4) gives:

$$Q_{W0} = mqN_{laser}(T_p) \tag{6}$$

$$Q_{W1} = \frac{z}{z_{\max}} mqN_{laser}(T_p) \tag{7}$$

These values are affected by the photon shot noise [4]; in particular, the background contribution to the noise is not null and has an equivalent integration time of $4T_p$, obtaining:

$$\sigma_{QW0} = q\sqrt{m\left(qN_{laser}(T_p) + qN_{bg}(4T_p)\right)} \tag{8}$$

$$\sigma_{QW1} = q\sqrt{m\left(\frac{z}{z_{\max}} qN_{laser}(T_p) + qN_{bg}(4T_p)\right)} \tag{9}$$

Using the known formula for the ratio error propagation, the distance uncertainty is obtained, taking into account only quantum noise of generated electrons:

$$\sigma_{zTOF} = z_{\max} \frac{Q_{W1}}{Q_{W0}} \sqrt{\left(\frac{\sigma_{QW0}}{Q_{W0}}\right)^2 + \left(\frac{\sigma_{QW1}}{Q_{W1}}\right)^2} \tag{10}$$

This equation allows calculating the shot noise limit of a pulsed light I-TOF imager, and also to evaluate the effect of the background light on the sensor precision, assuming that the cancellation is perfect. However, electronics noise typically dominates in this kind of 3D imager and cannot be neglected: on the other hand, the electronics noise strongly depends on the actual implementation of the sensor.

2 Case Study 1: 50 × 30-Pixel Array with Fully Differential 3D Pixel

The sensor described in [1], called "3DEye", is a 3D image sensor of 50 × 30 pixels and implements a fully differential structure in the pixel to realize the operations previously described in Fig. 1.

2.1 Sensor Design

The 3DEye pixel can be described with the picture of Fig. 2. The input is provided by two photodiodes, where one is active and receives the incoming light, while the other is a blind dummy to keep the symmetry of the circuit.

Applying the results of the previous paragraph to the 3DEye pixel, the INT signal determines the observation window allowing the photogenerated current to be integrated onto the capacitors and the chopper at the input allows cancellation of the background signal through integration of the input with alternating sign.

Due to the relatively large time between accumulations (a single accumulation lasts only hundreds of nanoseconds), to avoid common mode drift at the input, as well as undesired charge integration, the INT switches are opened and the RES switches are closed. The RES switches are closed together with INT only once, at the beginning of the accumulation.

The architecture of Fig. 2 has some clear advantages that can be summarized in the following list:

- The circuit is simple and clean and is easy to analyze
- The fully-differential topology guarantees reliability and low offsets
- The background cancellation is very effective

On the other side, there are some drawbacks that must be considered before scaling the architecture:

- The area used by the dummy photodiode is somehow "wasted"
- The noise is quite high due to kTC that adds each accumulation
- The output swing is limited due to undesired charge sharing and injection

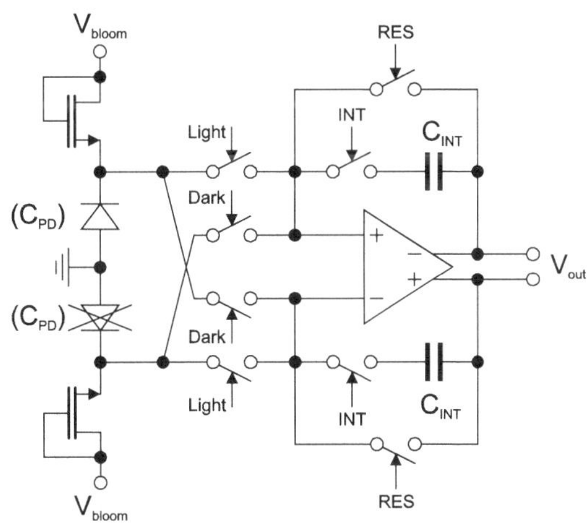

Fig. 2 Principle of operation of the 3DEye pixel

Table 1 3DEye measurement parameters

Parameter	Value
P_{laser}	240 W
T_p	30 ns
θ_{laser}	15 deg
F#	1.4
QE@λ_{laser}	0.2@900 nm
QE@λ_{bg}	0.3@550 nm
τ_{opt}	0.9
ρ	0.9
A_{pix}	1600 μm^2
FF	20 %
C_{INT}	56 fF
C_{PD}	160 fF

The data that will be used in the calculations is referred to the measurement conditions of the 3DEye demonstrator and is summarized in Table 1.

2.1.1 Theoretical Limit

The graphs of Fig. 3 show the absolute and relative error of the measured distance, with 32 accumulations, in the specific case 20 fps operation, with the data of Table 1 and Eq. (10). Each chart presents the precision without background light, and with 50 klx of background light. The integration of each component has been considered to be respectively $4T_p$, T_p, $T_p \cdot z/z_{max}$, as stated before.

As it can be seen, the error increases with the distance due to the weakening of the reflected laser pulse; moreover the background acts as a noise source, worsening the performance.

2.2 Circuit Analysis

The main source of uncertainty in the circuit of Fig. 2 is the kTC noise of the switches [5]. When a switch opens on a capacitor, it samples its thermal noise onto that capacitor which results to be independent from the switch resistance and with a standard deviation of $\sqrt{kT/C}$: to obtain the noise expression, each switch must be considered individually.

The switches of the chopper exchange the sign four times each measurement, but if this happens when the RES is active, their contribution is not added.

The reset switch acts during the initial reset sampling its noise on the integration capacitances, and during its opening before each accumulation. The total noise at the output due to reset switch is:

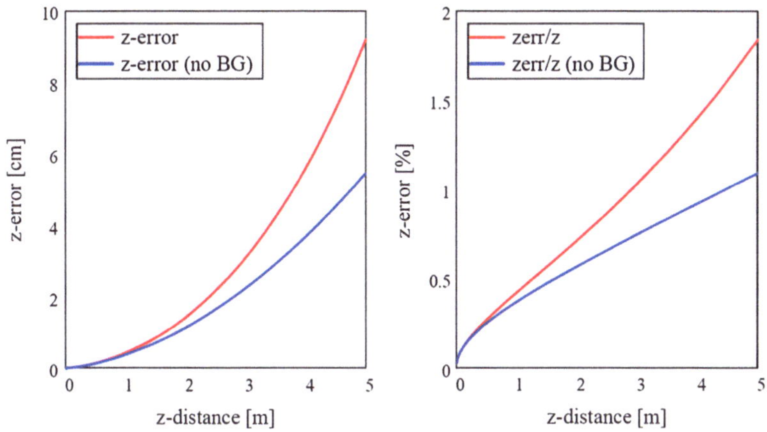

Fig. 3 Absolute and relative error with 32 accumulations (20 fps)

$$V^2_{kTC-RES} = 2\left(\frac{kT}{C_{INT}} + 2m\frac{C^2_{PD}}{C^2_{INT}}\frac{kT}{C_{PD}}\right) = \left(2 + 4m\frac{C_{PD}}{C_{INT}}\right)\frac{kT}{C_{INT}} \tag{11}$$

The integration switches open twice per accumulation (laser and background), and sample on one side onto the integration capacitances and on the other onto the photodiodes; the latter is then cancelled by the RES switch. After the last accumulation the INT switch remains closed while all the chopper switches open, in such a way that it is possible to read-out the pixel. The final noise due to integration switch is:

$$V^2_{kTC-INT} = 2(2m-1)\frac{kT}{C_{INT}} \tag{12}$$

Another contribution is the thermal noise of the operational amplifier that in the same way of the kTC is sampled and added to the total output noise during each release of the INT switch. So the fully-differential OTA noise (see [6]) can be expressed as:

$$V^2_{OTA} = 2(2m-1)\frac{C_{PD} + C_{INT}}{C_{INT}}\frac{kT}{C_{INT}} \tag{13}$$

2.3 Measurements and Outlook

Using Eqs. (6) and (7) for the calculation of the integrated charge during the two windows, Eqs. (8) and (9) for the shot noise, and Eqs. (11–13) for the electronics noise, it is possible to obtain all the necessary data in the case of medium illumination (50 lx or 0.073 W/m^2, typical living room condition).

Table 2 Calculated and measured values for [1]

	z = 1 m	z = 2 m	z = 3 m	z = 4 m
V_{W0}	1140 mV	284 mV	126 mV	71 mV
σ_{VW0}	1.35 mV	0.67 mV	0.45 mV	0.33 mV
V_{W1}	153 mV	126 mV	84 mV	63 mV
σ_{VW1}	0.64 mV	0.45 mV	0.37 mV	0.32 mV
σ_{Vel}	8.51 mV	8.51 mV	8.51 mV	8.51 mV
σ_{zcalc}	4 cm	15 cm	37 cm	73 cm
σ_{zmeas}	6 cm	20 cm	44 cm	82 cm

At the same time, it is possible to measure and compare the actual and expected values at different distances.

As it can be seen from Table 2, the noise amplitude due to the quantum nature of electrons is lower than the electronics noise. Taking into account all noise sources and using the formula for the error propagation, the predicted distance uncertainty results in very good agreement with the measurements and shows a large electronics noise contribution.

Example of the operation of the sensor in a 3D imaging system can be seen in Fig. 4, where sensor output of a hand indicating number "3" in front of a reference plane is shown. Both grayscale back-reflected light and color-coded 3D are given.

2.3.1 Improving the Sensor

The analytic tools developed in the last paragraphs allow prediction of future improvements of the 3DEye architecture by exploiting scaling with deep submicron technologies.

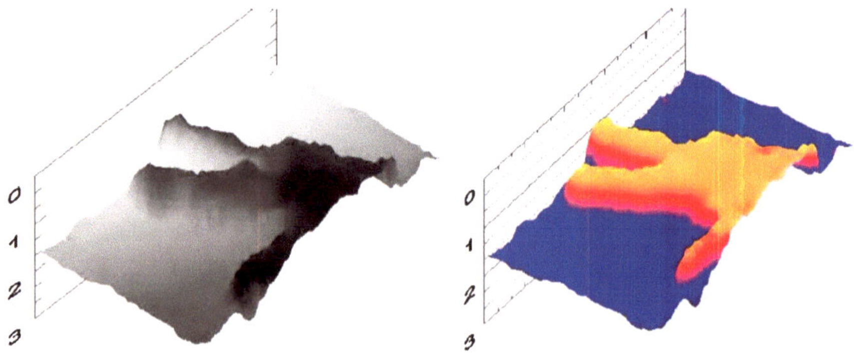

Fig. 4 Hand in both intensity and color-coded 3D mode captured with the 50 × 30-pixel fully-differential sensor

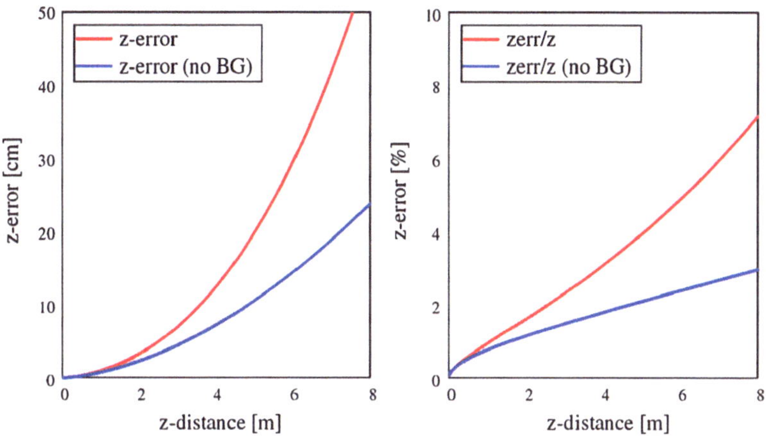

Fig. 5 Scaled pixel limits (240 W-50 ns pulse, 30 μm pitch, 30 % fill-factor, 32 accumulations)

Table 3 Voltages for the scaled pixel

	z = 4 m	z = 6 m
V_{W0}	140.2 mV	62.3 mV
σ_{VW0}	1.5 mV	1.0 mV
V_{W1}	74.8 mV	49.9 mV
σ_{VW1}	1.1 mV	0.9 mV
σ_{Vel}	29.2 mV	29.2 mV
σ_z	1.77 m	4.52 m

Starting with the optimistic hypothesis of the feasibility of the same electronics in 30 μm pitch with 30 % fill factor, the new performances of the hypothetic scaled sensor can be calculated, with enhanced range up to 7.5 m (50 ns laser pulse).

In Fig. 5 the forecast of the performance limit is plotted against the distance: it can be seen that the reduction of the pixel size strongly affects the precision.

With an integration capacitance C_{INT} of 10 fF (for a 0.18 μm technology it is almost the minimum that can be reliably implemented), and an estimated photodiode capacitance of 70 fF, also the electronics noise can be taken into account. The integrated voltage and the shot noise in the case of medium illumination (50 lx or 0.073 W/m^2, typical living room condition) and for two different distances is shown in Table 3.

The obtained numbers tell that the improvements in fill-factor and integration capacitance are not enough to compensate for the weaker signal. Therefore, it is clear that a scaling of the pixel is only possible with a different architecture.

3 Case Study 2: 160 × 120-Pixel Array with Two-Stage Pixel

From the analysis of the previous paragraph, it becomes clear that the main point is to reduce the electronics noise contribution. This can be achieved by pursuing several objectives:

- Increase the number of accumulations, and speed-up the readout of the sensor
- Maximize the fill-factor and the gain of the integrator
- Perform pixel-level correlated double sampling (CDS) to remove reset noise

3.1 Sensor Design

In order to accumulate the integrated value, without repeatedly sampling and summing noise on a small integration capacitance, the value must be transferred from the integrator to an accumulator stage using a larger capacitance. Such a circuit can be implemented in a single ended fashion, simplifying the complexity and optimizing the SNR, and can also be used as a CDS to remove most of the integrator's noise contribution.

The resulting two-stage architecture can be seen in Fig. 6: the two stages have a reset signal, and are connected by a switch network. The network operates in three modes that are identified by the control signals:

- INT: the integrator is connected to C_1 which is charged to V_{INT}
- FWD: C_1 is connected to the 2nd stage and summed to the value in C_2
- BWD: C_1 is connected to the 2nd stage and subtracted to the value in C_2

To achieve the I-TOF behavior the phases should be arranged in such a way that the difference between the end and start of the observation windows is calculated: this can be done also including the CDS for the first stage.

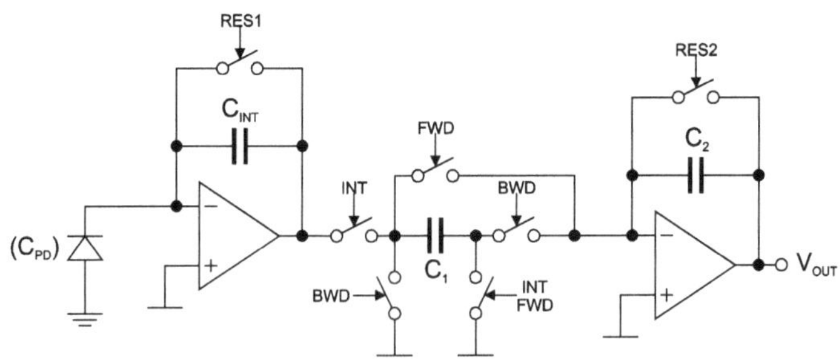

Fig. 6 Schematic representing the two-stage 3D pixel

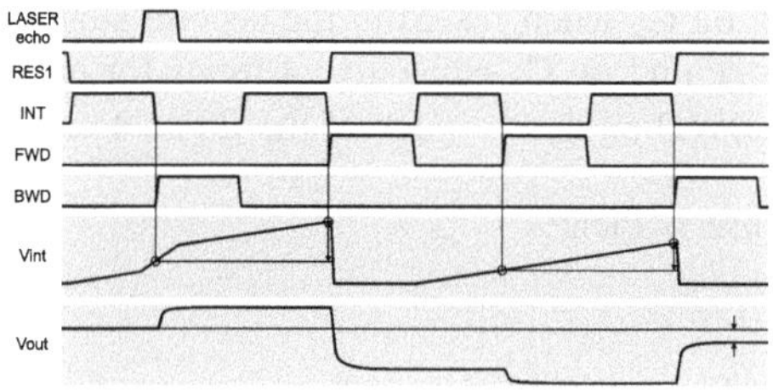

Fig. 7 Waveforms for the in-pixel pulses accumulation

As Fig. 7 shows, with proper arrangement of FWD and BWD it is possible to perform a true CDS and background light removal: the waveforms can be repeated without resetting the second stage, so accumulating the signal on V_{OUT}. The result is that only the portion of laser pulse that arrives between the first and second falling edges of the INT signal is integrated and summed to the output.

Figure 8 shows the block schematic of the sensor architecture, which includes logic for addressing, column amplifiers for fast sampling and readout, and driving electronics for digital and analog signal distribution.

3.2 Circuit Analysis

The same noise analysis done on the fully-differential pixel can be performed for the two-stage pixel. The contributions to be considered are the first and second stage kTC noise, and the first and second stage OTA noise. Thanks to the CDS operation, the first stage kTC noise is cancelled, so only the second stage kTC noise is relevant. There are three terms in the second stage kTC noise:

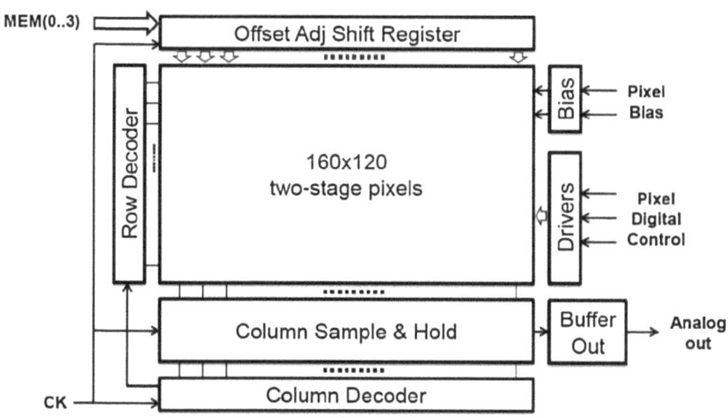

Fig. 8 Architecture of the 3D image sensor

- one-shot sampling on C_2 when resetting the second stage
- kTC due to INT opening on C_1 capacitor, 4 times for each accumulation
- kTC due to FWD/BWD on the "-"node of the OTA, 4 × for each accumulation

The total kTC noise is then:

$$V_{kTC1}^2 = 0$$
$$V_{kTC2}^2 = \frac{kT}{C_2} + 4m\frac{C_1^2}{C_2^2}\frac{kT}{C_1} + 4m\frac{kT}{C_2} = \left[1 + 4m\left(1 + \frac{C_1}{C_2}\right)\right]\frac{kT}{C_2} \tag{14}$$

As far as the OTA noise is of concern, in the first stage it is given by the OTA noise sampled onto the C_1 capacitor at the end of each INT signal, while the OTA noise sampled during the reset is removed by the CDS. The equation takes into account a single-input amplifier and the transfer function to the output C_1/C_2. The second stage samples the OTA noise once during reset, then at FWD/BWD opening, 4 times for each accumulation.

$$V_{OTA1}^2 = 4mkT\frac{C_{PD} + C_{INT}}{C_{INT}(C_{INT} + C_1)}\left(\frac{C_1}{C_2}\right)^2$$
$$V_{OTA2}^2 = 2(1 + 4m)\frac{C_1 + C_2}{C_2}\frac{kT}{C_2} \tag{15}$$

3.2.1 Two-Stage Pixel Simulation

The implementation of the pixel uses the following capacitances of $C_{INT} = 9.5$ fF, $C_1 = 65$ fF, $C_2 = 130$ fF, while the estimated photodiode capacitance, at 0.9 V bias results $C_{PD} = 80$ fF. A complete distance measurement simulation can be carried out, with the conditions of Table 4, which refer to the final testbench for the sensor.

Table 4 Parameters used for simulation of measurement

Parameter	Value
P_{laser}	250 W
T_p	50 ns
θ_{laser}	15 deg
F#	1.4
QE@λ_{laser}	0.2@900 nm
QE@λ_{bg}	0.3@550 nm
τ_{opt}	0.9
ρ	0.9
A_{pix}	846.8 μm^2
FF	34 %
C_{INT}	9.5 fF
C_{PD}	80 fF

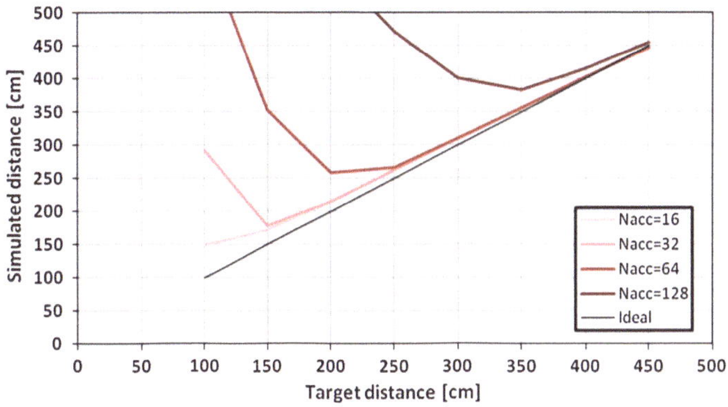

Fig. 9 Simulated distance with 16, 23, 64, 128 accumulations

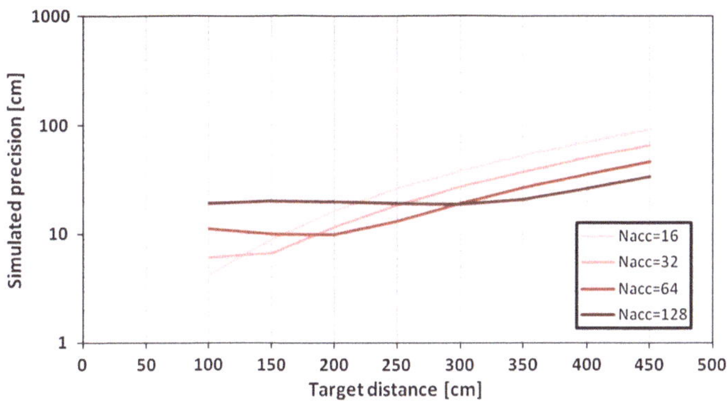

Fig. 10 Simulated precision from calculated noise and simulated voltages

Using the parameter of Table 4, together with the calculations at the beginning of the chapter and Eqs. (14) and (15), it is possible to extract some performance forecast. In Fig. 9 the calculated distance, compared with the ideal one, is plotted: at small distances the higher number of accumulations saturates (W_0 saturates first), while the line has a smaller slope with respect to the ideal characteristics due to the non-ideal response of the integrator in time. In Fig. 10 the precision of the distance measurement is extracted.

3.3 Test Setup and Measurement Results

The sensor, implemented in a CMOS 0.18 μm technology, is visible in Fig. 11: a detail of the pixels is shown in the inset.

The size of the chip is 5 × 5 mm^2 and it contains 160 × 120 pixels with pitch of 29.1 μm implementing the circuit of Fig. 6.

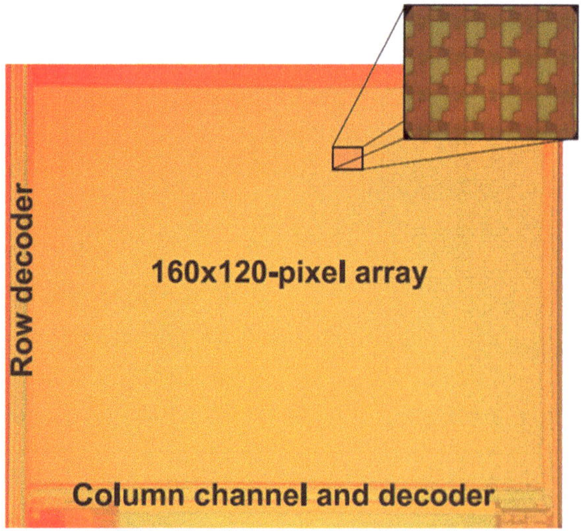

Fig. 11 Chip micrograph of the 160 × 120-pixels 3D sensor

3.3.1 Description of Setup

The camera board has to be mounted on an optical bench; the 905 nm laser source is installed immediately above the optics to ensure good matching between excitation and captured image, as visible in Fig. 12: it allows 50 ns-wide pulses at a maximum duty-cycle of 0.1 %. Each of the three modules has a peak power of 75 W, for a total peak power of 225 W.

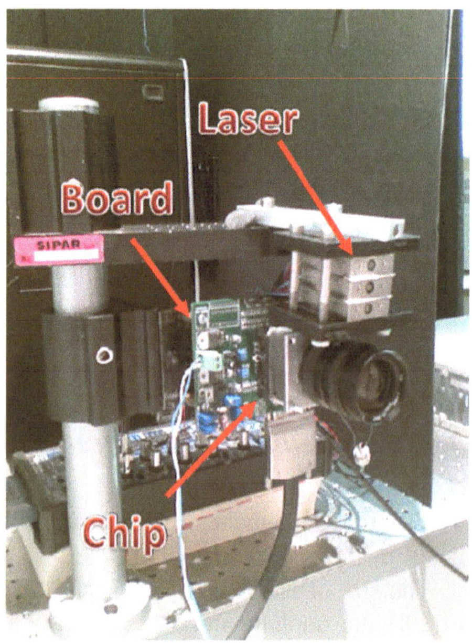

Fig. 12 Photograph of the setup on the optical bench

A large panel white panel with estimated reflectivity of 40 % is used as a target and placed in front of the camera-laser system.

3.3.2 Distance Measurements

The charts of Figs. 13 and 14 show the measured distance and the achievable precision, respectively, for different number of accumulations. In particular, from 16 to 128 accumulations the frame rate results to be of 65.0 fps down to 27.8 fps, with an average illuminator power from 26 to 89 mW.

The measurements compare well with the estimation of previous paragraphs, taking into account the reduced reflectivity of the target employed in the experimental conditions.

A scene showing a teddy bear and an apple, acquired averaging several frames for a total acquisition time of 7 s, is shown in Fig. 15.

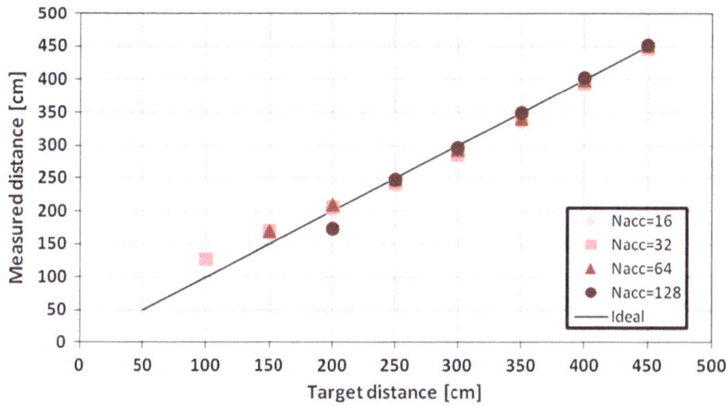

Fig. 13 Measured distance for 16, 32, 64, 128 accumulations

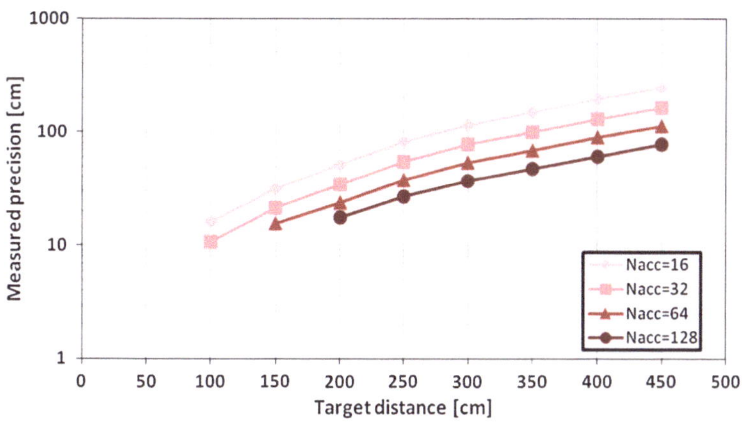

Fig. 14 Measured precision for 16, 32, 64, 128 accumulations

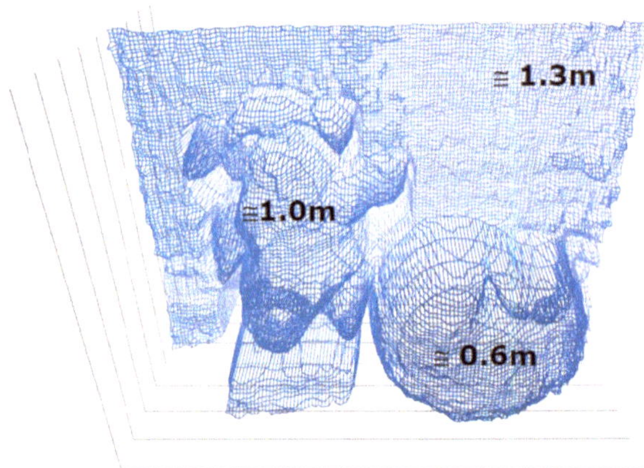

Fig. 15 Scene composed by teddy bear, apple and background at different distances

4 Correlation I-TOF 3D Imaging Method

The time of flight correlation method implies that an optical signal with phase shifted rectangular sequence modulation illuminates the scenery, whereby a small part of it reflects back to the sensor. The backscattered light is subsequently in each pixel correlated with internally generated signal, which is also of rectangular shape but with fixed phase. The correlation function of these two (presumably rectangular) signals is of a triangular shape, formed by N discrete phase steps that are applied to the illumination source. Due to the usage of digital devices for the signal generation, it makes sense for N to be a power of two, and is here chosen to be $N = 16$. The continuous-time correlation function is now defined as

$$CF(\tau) = \int s_{REC}(t) s_{CLK}(t + \tau) dt \tag{16}$$

where s_{REC} represents received optical power, and s_{CLK} internal generated signal. In order to sample the correlation triangle at N points, the phase of the modulation signal that controls the light source is shifted throughout equally spaced phase steps so that $s_{MOD}(t) = s_{CLK}\left(t - \frac{n}{Nf_{MOD}}\right)$ with n as a phase step counter. The correlation process (multiplication and integration) is performed in each pixel separately and will be discussed in detail in the next section.

For the calculation of the distance information, the phase displacement information φ_{TOF} is essential. The extraction of φ_{TOF} is done by applying a discrete Fourier transform (DFT) on the correlation triangle and recovering thereby the phase of the fundamental wave. Figure 16 depicts the control signal timings as well as the differential pixel output ΔV_{OUT}, which is the correlation triangle function.

The corresponding fundamental wave of the correlation triangle is sketched additionally. All control signals shown in Fig. 16 are generated and controlled by means of an FPGA and applied to the illumination source and to the sensor as shown in Fig. 17. Each phase step begins with a *reset* for providing the same initial

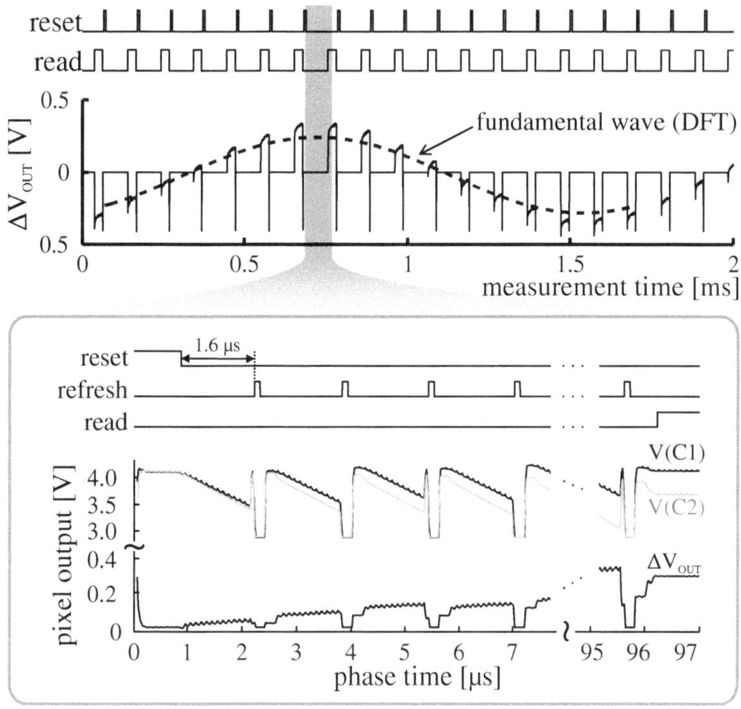

Fig. 16 Signal timing with corresponding pixel output of a complete correlation triangle and one phase step in detail

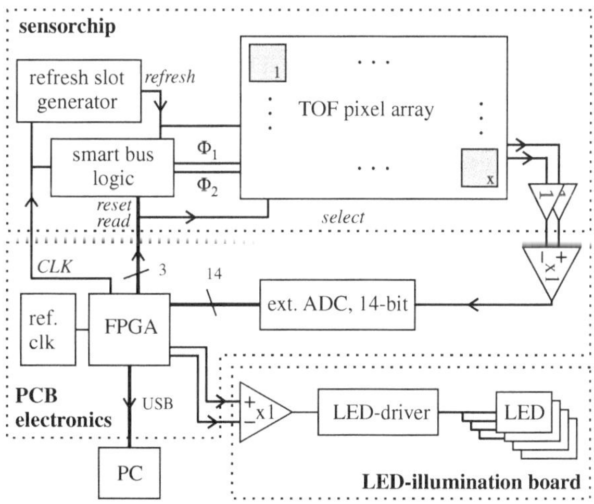

Fig. 17 Sensor block diagram

conditions prior to every integration process. After the reset process is completed, an integration interval takes place defining the differential voltage. Eventually, a *read* cycle is applied for the read out purpose. The corresponding phase step output voltage of the sensor is thereby only applied to the output during the *read* cycle. An additional control signal (*refresh*) that eliminates the extraneous light contributions interrupts the integration process after every 1.6 μs consequently providing initial common mode voltages (see pixel output signals in Fig. 16).

4.1 Sensors Architecture

The full optical distance measurement system consists of three main components, which are located on separated PCBs: the sensor itself as an optoelectronic integrated circuit (OEIC); the illumination source, which is implemented as a red laser for single pixel characterisation and as a LED source for full sensor characterisation; and an FPGA PCB for control and signal processing. Figure 17 shows a block diagram of these units and the interconnection setup. As mentioned in the previous section, the sent optical signal must undergo a phase shift of 16 phase steps in order to obtain a correlation triangle, whereby, the signal that is applied to the correlation circuit in each pixel remains the same for all 16 phase steps. Previous results based on this approach used a bridge correlator circuit that served for the background light extinction (BGLX) as reported in [7] or an operational amplifier in [8]. Both designs suffered from relatively high power consumption and in the later approach also large pixel area, amounting to 109 × 241 μm^2. In the following, a radical improvement in terms of background light suppression capability, as well as power consumption will be presented.

This section describes two TOF pixel sensors reported in [9] and [10] with different implementations of BGLX techniques. The first approach, consisting of 2 × 32 pixels configured in a dual line architecture, is described in Sect. 5. In Sect. 6, another technique for BGLX comprising 16 × 16 pixel array is shown. Both sensors are fabricated in a 0.6 μm BiCMOS technology including photodiodes with anti-reflection coating. However, in both chips only CMOS devices are exploited for the circuit design.

4.2 Signal and Noise Analysis

The received optical signal is converted by the photodiode into a photocurrent in the TOF sensor. The hereby generated current I_{MOD} is a function of the photodiode's responsivity R and the light power P_{pix} of the received optical signal. The relationship between the received light power P_{pix} at the pixel and the transmitted light power P_{laser} is

$$\frac{P_{pix}}{P_{laser}} \propto \rho \frac{w^2}{F\#^2 z_{TOF}^2} FF \tag{17}$$

taking into account that any atmospheric attenuation was neglected. In Eq. (17) ρ is the reflection property of the target, w is the edge length of the squared pixels, z_{TOF} is the distance between pixel and object and FF is the optical pixel fill factor. Considering that a responsivity R amounts to 0.33 A/W at 850 nm and received light power P_{pix} is in the nanowatts range, a photocurrent I_{MOD} in the nanoampere range can be expected. This current is accompanied by the photon noise current $i_{MOD} = \sqrt{qI_{MOD}/T_{\Delta}}$, where q is the electron charge and T_{Δ} is the effective integration time in a complete correlation triangle covering all 16 phase steps. The noise current I_{MOD} is the limiting factor of optical systems which is caused by photon noise [11]. Further noise components based on background light i_{BGL}, bias current i_0, A/D-conversation $i_{A/D}$, switching process during correlation (kT/C) $i_{kT/C}$ and other electrical noise contributions i_{EL} like output buffer noise, substrate noise, power supply noise, can be transferred to the input of the circuit. At the cathode of the photodiode D_{PD} a signal-to-noise ratio (SNR) of a single correlation triangle SNR_{Δ} can be given by

$$SNR_{\Delta} = \frac{I_{MOD}^2}{i_{MOD}^2 + i_{BGL}^2 + i_0^2 + i_{A/D}^2 + i_{kT/C}^2 + i_{EL}^2} \approx \frac{I_{MOD}^2}{i_0^2 + i_{kT/C}^2} \tag{18}$$

The dominating parts of the upper equation are the kT/C-noise and the noise current due to the bias which is expected to be in the same order of magnitude as the switching noise. The measurement accuracy is primarily determined by the bias current I_0, since this current is about 5 to 10 times higher than the current due to background light illumination I_{BGL}. It should be noted that Eq. (18) is an implicit function of the total time spent for integration T_{Δ} behind the noise components. However, this equation is absolutely independent of the number of the phase steps within T_{Δ}. The standard deviation for distance measurements is given by

$$\sigma_{zTOF} = \frac{\sqrt{2}\pi}{32} \frac{c_0}{f_{MOD}} \frac{1}{\sqrt{SNR_{\Delta}}} \tag{19}$$

This equation is based on SNR_{Δ} as the characteristic number for the influence of stochastic sources of errors. For non-perfect rectangular modulation the prefactor in Eq. (19) is not valid and has to be adapted. By means of this equation measurement accuracy at a certain distance can be predicted in advance.

5 Case Study 3: 2 × 32 Dual-Line TOF Sensor

In this subsection a dual line TOF sensor with a 2 × 32 pixel array and a total chip area of 6.5 mm^2 (see Fig. 18) is presented. Additional parts like output buffers for driving the measurement equipment, a logic unit for processing the smart bus signals and a phase-locked-loop (PLL) based ring-oscillator for optional generation of the shifted modulation signals on-chip in combination with a 50 Ω-driver

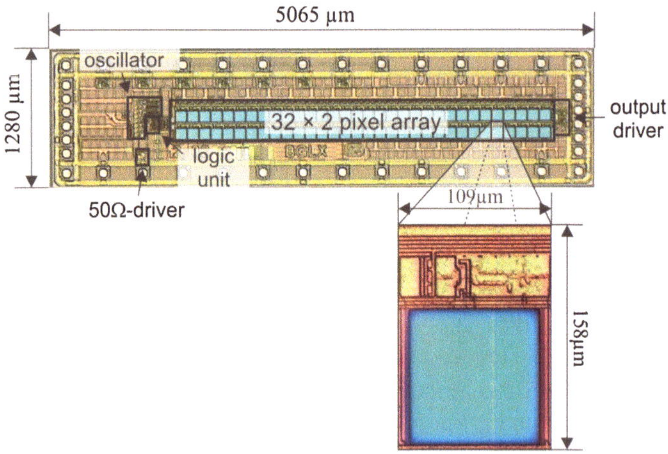

Fig. 18 Chip micrograph of the 2 × 32 pixel sensor

dedicated to drive the illumination source, are included in the chip. The required signals for the chip can be internally generated by the PLL or in an FPGA. The main drawback of using the internally generated signals is that substrate and power supply noise affect the performance of the sensor. Each pixel covers an area of $109 \times 158\ \mu m^2$ including an optical active area size of $100 \times 100\ \mu m^2$, which results in an fill-factor of $\sim$58 %. The pixel consumes 100 μA at a supply voltage of 5 V, whereby, the internal OPA dominates the pixel power consumption. The power consumption of the whole chip is nearly 80 mW, whereof 92 % are used by the output buffers and the on-chip phase generator.

5.1 Sensor Circuit

The circuit of a single pixel used in the dual line sensor is shown in Fig. 19. It features fully differential correlation according to the underlying double-correlator concept presented in [12]. The received photocurrent I_{PH} is directed through transistors T1 and T2 to the corresponding integration capacitors C_{I1} and C_{I2}. Dummy transistors T1d and T2d were added in the path to compensate charge which is injected during the integration interval. Two memory capacitors C_{M1} and C_{M2} with the transistors T3–T9 form the circuit part for the BGLX. The BGLX process is controlled by the $\overline{bglx}$ signal, which is deduced from the *refresh* signal. During the read process the voltages at the integration capacitors C_{I1} and C_{I2} are buffered to the output by activating the *select* signal. Afterwards, throughout the readout process, the OPA output voltage has to be defined. This is done by transistor T10 and a XNOR logic gate. Transistor T10 and the XNOR are also used during the

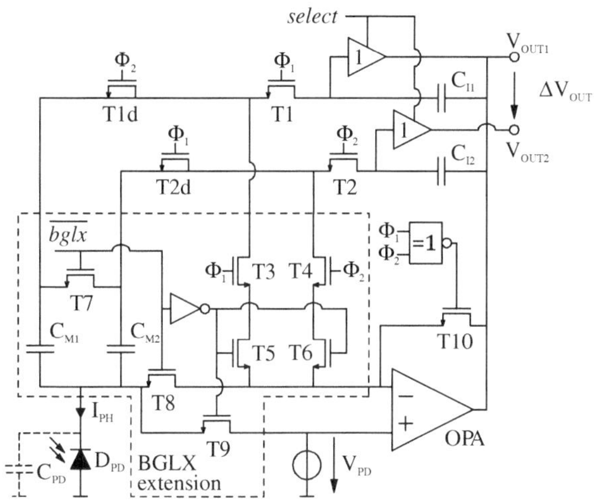

Fig. 19 Pixel circuit with sketched part for background light suppression of the 2 × 32 pixel sensor

reset cycle, where they provide a short to the integration capacitors. The 2-stage OPA has to provide an output swing of ±1.5 V with a fast slew rate for regulating the cathode voltage of the photodiode D_{PD} according to V_{PD} at the positive input. A gain of 68 dB and a transit frequency of 100 MHz of the OPA are adequate for achieving the requested regulation.

Before starting with the integration, a reset is applied to the circuit by forcing Φ_1 and Φ_2 to high and $\overline{bglx}$ to low. Thereby C_{I1} and C_{M1} as well as C_{I2} and C_{M2} are cleared via T1–T6 in combination with activating T9 and T10. After the circuit reset is performed, the accumulation of the photocurrent I_{PH} starts by setting $\Phi_1 = \overline{\Phi_2}$ and $\overline{bglx}$ to high. Depending on the level of Φ_1 and Φ_2 and thus on the modulation clock, the photogenerated current is collected either in C_{I1} or C_{I2}. The two memory capacitors C_{M1} and C_{M2} are switched during the integration in parallel. Hence, the half of photogenerated charge is stored in them, presupposing that all four capacitors are of the same size. Thereby the integrated photocurrent I_{PH} consists of two parts due to intrinsic modulated light I_{MOD} and the background light I_{BGL}. The charge stored in each memory capacitor is $(I_{BGL} + I_{MOD})/2$, while the charge stored in C_{I1} and C_{I2} is an identical amount of $I_{BGL}/2$ and the portion of I_{MOD} due to the correlation.

BGLX technique is necessary since the current due to background light I_{BGL} is in the μA range for light conditions of around 100 klx and thus more than 1000 times larger than the part of the modulated light I_{MOD}, which is in the nA or even pA range. The BGLX process is provided in periodic refresh intervals during one system clock period. Thereby the $\overline{bglx}$ signal is set to low, while the modulation signals Φ_1 and Φ_2 continue with operation as during the integration period. The common node of D_{PD}, C_{M1} and C_{M2} is connected to V_{PD} due to the low state of the $\overline{bglx}$ signal. As in the integration phase transistors T1 and T2 continue passing the photocurrent to the integration capacitors C_{I1} and C_{I2} during the half-cycles.

Additionally transistors T3 and T4 are controlled with the same clock like T1 and T2. The load of the capacitors C_{M1} and C_{M2} is sensed by the OPA through the low-ohmic switching transistors T5 and T6. Afterwards the OPA compensates with I_{COMP} for the charge applied to its differential input. This process leads to subtraction of charges due to ambient light in the integration capacitors C_{I1} and C_{I2}. Since each memory capacitor is loaded with a charge proportional to $I_{BGL}/2 + I_{MOD}/2$, only a part originating from $I_{MOD}/2$ is subtracted from the integration capacitors. This involves a common-mode-corrected integration at both capacitors after each BGLX cycle, enabling a background light suppression up to 150 klx.

5.2 Test Setup and Measurement Results

As depicted in Fig. 17 the setup for characterizing the sensor performance consists of three components. The illumination PCB is mounted on a small aluminium breadboard in front of the PCB carrying the TOF sensor. For the dual line pixel sensor a 1-inch lens with a focal length of 16 mm and an F number of 1.4 was used. Since a strong background light source for illuminating a complete sensor's field of view up to 150 klx was not available, single pixel measurements were done for characterizing the pixel under high background light conditions. In this case a collimated light from a red laser at 650 nm with an average optical output power of 1 mW was employed as an illumination source. Furthermore the usage of a laser spot illumination would be of advantage due to a high modulation bandwidth of the laser diode. The applied modulation frequency was $f_{MOD} = 10$ MHz, leading to a non-ambiguity range of 15 m. For measurements with the dual line TOF pixel sensor a field of illumination according to the sensor's field of view (FOV) is necessary. Therefore, a laminar illuminated FOV is provided by a modulated LED source with a total optical power of nearly 900 mW at 850 nm. The LED source consists of four high-power LED devices of type SFH4230 in combination with 6° collimator lenses each, whereby each LED is supplied with 1 A at 10 MHz and emits an optical power of 220 mW. Each LED has its own driver circuit, consisting of a buffer which drives the capacitive load of the gate of an NMOS switching transistor with a low on-resistance and a serial resistance for operating point stabilisation. Due to the angular field of illumination, this illumination source has a lower light power compared to the laser spot illumination source used for the characterisation of a single pixel. By means of the 6° collimator lenses an optical power density of 8 W/m^2 is achieved at a distance of 1 m. Therefore, eye safety conditions are achieved for distances $d > 10$ cm. A circular cut-out between the LED pairs was made for mounting a lens system along the optical axis. Furthermore, a 50 Ω terminated differential input converts the modulation clock to a single-ended signal for the level converter, which provides the signal for the following LED driver circuit. The modulation clock for the illumination source is provided by the FPGA.

As an object, a 10 × 10 cm^2 non-cooperative white paper target with about 90 % reflectivity was used for characterization in single-pixel measurements. Thereby the target was moved by a motorized trolley with a linear range of 0.1 to 3.2 m. The upper range is limited by the laboratory room and thus the length of the linear axis. For characterizing the dual line TOF pixel sensor a moving white paper target and a laboratory scenery where used, respectively. The robustness to extraneous light was verified by a cold-light source with a colour temperature of 3200 K that illuminated the white paper target with up to 150 klx. After the digitalization of the correlation triangle the digitalized data pass the DFT-block in the FPGA, whereby the amplitude and phase of the fundamental wave are extracted. Once the phase acquisition is done, the distance can be calculated presupposing known measurement conditions as explained in [12].

5.2.1 Single Pixel Characterization

For characterizing a single pixel, measurements with 100 measured distance points are recorded at a step size of 10 cm, while the measurement time for a single point was 50 ms. The collected data were analysed and the standard deviation σ_{zTOF} as well as the linearity error e_{lin} were obtained. Figure 20a depicts the results of these measurements. The accuracy is deformed in the near field due to defocusing of the sensor. Excluding this region, a standard deviation of a few centimetres and a linearity error within ±1 cm is characteristic for this measurement range. The impact of background light, which leads to an about 1000 times larger photocurrent compared to the photocurrent due to the modulation light is depicted in Fig. 20b for measurements at a distance of 1 m. In this figure Δz is the relative displacement. The best result Δz = 10 cm for 150 klx is achieved by setting the BGLX period to 0.4 μs as depicted in the figure. Here, it should be again mentioned that no optical filters where used to supress the undesired light.

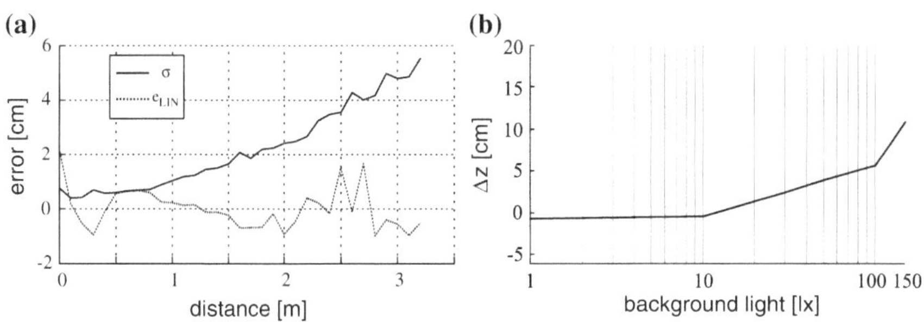

Fig. 20 Measurement results for the 2 × 32 sensor. **a** Error analysis of single pixel distance measurements up to 3.2 m, **b** Measurement error as a function of background light illumination intensity

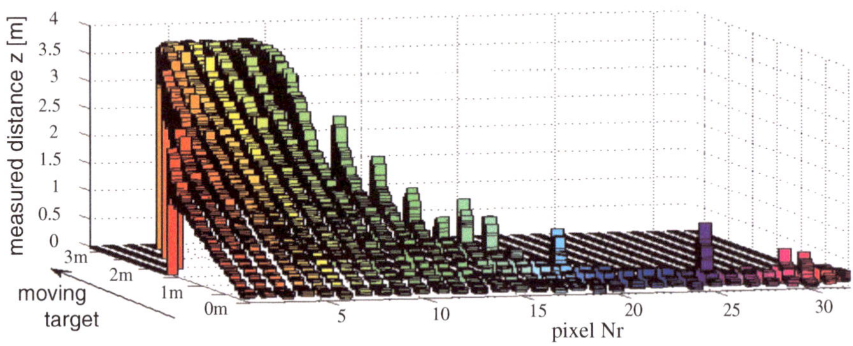

Fig. 21 Measurement results. Dual Line–Sensor measurement showing the movements of a white paper target

5.2.2 Dual Line Sensor Array Characterization

For the measurements of the multi pixel sensor the illumination source described in the corresponding section was used. The measurements show a standard deviation which is about 3.5 times higher than the single pixel results with the red laser as a direct consequence of lower reception power. Furthermore the linearity error also increases within a range of ±2 cm. Measurement results with the 2 × 32 pixel sensor are illustrated in Fig. 21, whereby the 10 × 10 cm^2 white paper target was moved back out of the sensor axis. For each position 100 records are sampled, whereby a total measurement time of 50 ms was used for each point. In this depiction, pixels with amplitudes below 1 mV were masked out. The amplitude value at a distance of 3.5 m corresponds to a current $I_{MOD} = 23$ pA.

6 Case Study 4: 16 × 16 Pixel Array TOF Sensor Design

In this subsection a 16 × 16 TOF sensor is presented. A micrograph of the chip is shown in Fig. 22. The sensor is based on another approach that can supress 150 klx, as well as the approach presented before. The whole chip covers an area of about 10 mm^2, whereby a single pixel consisting of the photodiode and the readout circuitry has a size of 125 × 125 μm^2 and an optical fill factor of ~66 %. Furthermore, some pixel logic, a reference pixel and a 12-bit analogue-to-digital converter are included in the chip. The A/D converter can be used for readout with 1 MS/s. The offset problem due to propagation delay of signals in the measurement setup can be fixed by means of the reference pixel, which is located outside the array. All 256 + 1 pixels are addressed by the *select* line. Thereby the readout mechanism is managed by a shift register, which is clocked by the *select* signal. This signal passes a "1" from pixel column to pixel column [13]. For digitalizing, the output voltage can be either converted by using the built-in ADC or by means

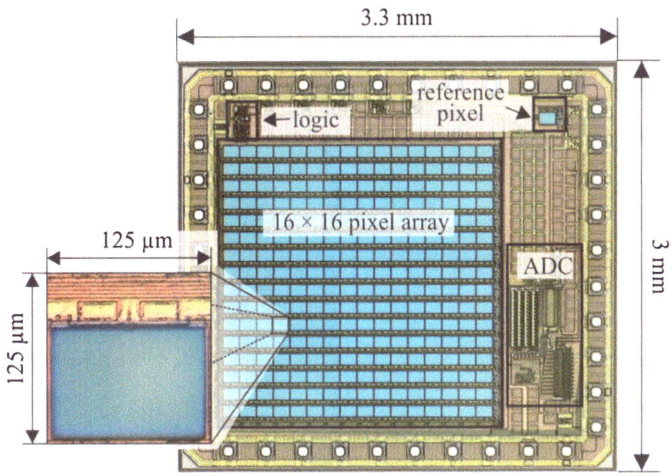

Fig. 22 Chip micrograph of the 16 × 16 pixel sensor

of an external ADC. Since no OPA is used for BGLX, the consumed current per pixel is 50 times smaller compared to the 2 × 32 pixel sensor presented above, amounting to only 2 μA at a supply voltage of 5 V. A disadvantage of the low current consumption is a decreased bandwidth of the pixel circuit. However, the bandwidth is still high enough to meet the TOF requirements at 10 MHz square-wave modulation.

6.1 Sensor Circuit

Figure 23 depicts the single pixel circuit. The pixel performs the correlation operation, which implies, similarly as in the first approach, a reset at the beginning of each phase step, an integration interval of several fundamental clock cycles afterwards, and readout at the end. The generated photocurrent from the photodiode D_{PD} is directed through transistors T6 and T7 to the integration capacitors C_1 and C_2. Both mentioned transistors are switched by the modulating clock signals Φ_1 and Φ_2 while the *refresh* signal is at low level. During the integration, a refresh process occurs periodically for removing the background light contribution that is stored onto capacitors C_1 and C_2. After 95 μs of repeated interchange of correlating integration and refresh activities, the differential voltages ΔV_{OUT} is read out. Thereby, the *read* signal forces Φ_1 and Φ_2 to ground which leads in directing I_{PH} over T5. With the pixel's *select* signal the output buffers are enabled and the differential output voltage ΔV_{OUT} is traced. During the integration process, transistors T10 and T11 regulate the voltage at the cathode of the photodiode to a constant value, suppressing thereby the influence of the photodiode capacitance C_{PD}. Transistor T10 needs to be biased with $I_0 = 1$ μA to ensure a sufficiently

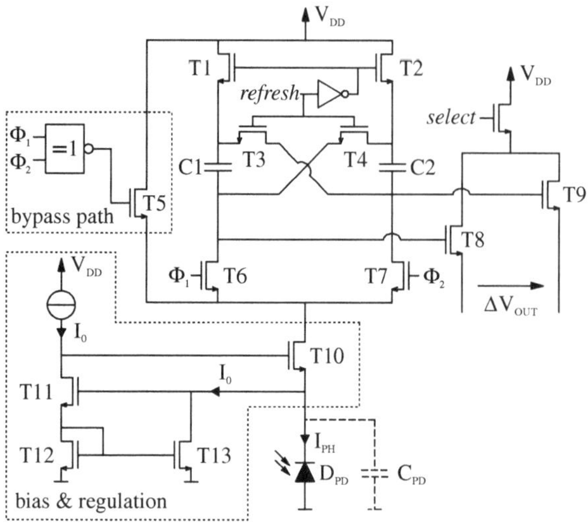

Fig. 23 Pixel circuit with sketched part for background light suppression of the 16 × 16 pixel sensor

large regulation bandwidth of 100 MHz for the 10 MHz square-wave modulation signals. This current is supplied by an in-pixel current source and mirrored by the transistors T11–T13. Since the current is mirrored, another 1 μA is drawn by the amplifying transistor T11 so that in total the pixel consumption can be kept at, as low as, 2 μA.

The BGLX operation is in this circuit ensured with the *refresh* signal. After some integration cycles *refresh* is activated to process the background light extinction. In addition to it, Φ_1 and Φ_2 are forced to ground. As a consequence, the transistor pairs T1 and T2 as well as T6 and T7 are switched in the high-ohmic region and isolate the integration capacitors. The still generated photocurrent I_{PH} is thereby bypassed over transistor T5. Transistors T3 and T4 force an anti-parallel connection of the integration capacitors due to the high state of the *refresh* signal, as described in [14]. Hence, the charge caused by background light is extinguished while keeping the differential information in each capacitor, which is depicted in Fig. 16. The refresh time interval is in this approach only half of a clock cycle of the double line sensor approach described earlier. After the background light extinction process the *reset* signal is forced to the low level and the integration can be continued.

6.2 Test Setup and Measurement Results

The test setup for the single pixel sensor was the same as presented already in Sect. 5.2. Also the test setup for the 16 × 16 TOF pixel array sensor was nearly the same as for the dual line TOF pixel sensor. The only difference was in the used

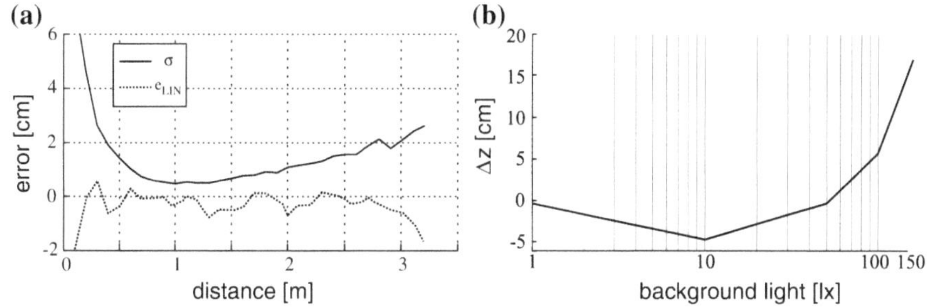

Fig. 24 Measurement results for the 16 × 16 sensor. **a** Error analysis of single pixel distance measurements up to 3.2 m, **b** Measurement error as a function of background light illumination intensity

lens for the light collection. Here, a commercial aspheric 0.5-inch lens with a focal ratio F $\approx$ 1 was used. The lens' focal length of 10 mm guarantees a blur free picture over the total measurement range from 0 to 3.2 m. Furthermore, for characterizing the 16 × 16 TOF pixel array sensor a laboratory scenery was used.

6.2.1 Single Pixel Characterization

Similarly as by the dual line pixel sensor, a single pixel characterization for the 16 × 16 array sensor was performed, exploiting the same 10 × 10 cm^2 white paper target under the same measurement conditions (100 measurements at each step and 10 cm steps). The results are depicted in Fig. 24a. A standard deviation σ_{zTOF} is below 1 cm up to 1 m and below 5 cm up to 3 m while the linearity error e_{lin} remains within –1/+2-cm band. The influence of background light at a distance of 1.5 m is depicted in Fig. 24b, achieving a displacement of $\Delta z = 5$ cm for background light of 100 klx and $\Delta z = 15$ cm for background light of 150 klx. The increase of Δz can be explained by a shift of the pixel's operating point due to the BGL-induced photocurrent.

6.2.2 Pixel Sensor Array Characterization

Laboratory scenery consisting of a black metal ring (at 0.4 m), a blue pen (at 0.6 m), a blue paper box (at 1.3 m) and the right arm of the depicted person which is wearing a red pullover at a distance of 2.2 m has been captured. Figure 25 clearly shows the 3D plot of this scenery and the colour coded distance information. The sensor chip is able to provide range images in real-time with 16 frames per second. Rough information about background light conditions can be extracted out of the common-mode information in the output signal, which could be used for correcting the effects on distance measurements in future implementations.

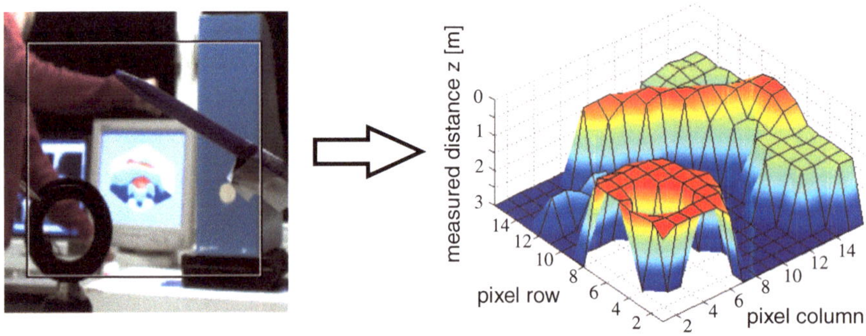

Fig. 25 Measurement results. 3D plot of a laboratory scenery measured by the 16 × 16 pixel array sensor

7 Discussion and Comparison

In this chapter, two techniques for the implementation of indirect Time-Of-Flight imaging sensors have been described in detail. Both methods adapt well to implementations making extensive use of electronics inside the pixel to perform extraction of the 3D information.

While the pulsed method allows obtaining a simple way to extract the 3D image, with a minimum of two windows measurement, the correlation method balances the need for more acquisitions and complex processing with a higher linearity of the distance characteristics. At the same time, the pulsed technique can perform background removal few hundreds of nanoseconds after light integration, while the correlation method performs the operation after several integration cycles, which may introduce bigger errors in case of bright moving objects. Both techniques allow achieving a very good rejection to background light signal, up to the value of 150 klux measured for the correlation method, which is a fundamental advantage of electronics-based 3D sensors.

Implementations of imaging sensors for both methods demonstrate the potential of the techniques to obtain moderate array resolutions, from 16 × 16 pixels up to 160 × 120 pixels at minimum pixel pitch of 29.1 μm. In all implementations, efforts in the optimization of electronics can bring to satisfactory fill-factors, and power consumption down to 2 μA per pixel in the case of the correlation-based pixel array. Integrated circuits are designed using different technologies without any special option, thus making this an attractive point for these techniques: no special process optimizations are needed, allowing easy scaling and porting of the circuits to the technology of choice.

Looking at the 3D camera system point of view, pulsed TOF requires the use of laser illuminator due to the high required power of the single pulse, while the correlation method has a relaxed requirement which allows the use of LEDs with a lower peak power. On the contrary, pulsed laser allows better signal-to-noise ratio thanks to the strong signal, while using modulated LEDs the signal amplitude becomes a very critical point, thus settling pixel pitch on 109 μm for the linear sensor and 125 μm for the array sensor. Anyway, both systems assess the average illuminator power requirements on similar orders of magnitude, and in such conditions both easily reach distance precision in the centimetres range.

The trend of electronics-based 3D sensors follows the reduction of pixel pitch and increase of resolution: it is likely to happen that these types of sensors will find applications where their strong points are needed, such as high operating frame-rate and high background suppression capability. Still several improvements can be done, like parallelization of acquisition phases and optimization of the circuit area occupation, which can bring to better distance precision. Therefore, applications in the field of robotics, production control, safety and surveillance can be envisaged, and future improvements could further enlarge this list.

References

1. R. Jeremias, W. Brockherde, G. Doemens, B. Hosticka, L. Listl, P. Mengel, A CMOS photosensor array for 3D imaging using pulsed laser, IEEE International Solid-State Circuits Conference (2001), pp. 252–253
2. D. Stoppa, L. Viarani, A. Simoni, L. Gonzo, M. Malfatti, G. Pedretti, A 50 × 30-Pixel CMOS Sensor for TOF-Based Real Time 3D Imaging, 2005 Workshop on Charge-Coupled Devices and Advanced Image Sensors (Karuizawa, Nagano, Japan, 2005)
3. M. Perenzoni, N. Massari, D. Stoppa, L. Pancheri, M. Malfatti, L. Gonzo, A 160×120-pixels range camera with in-pixel correlated double sampling and fixed-pattern noise correction. IEEE J. Solid-State Circuits **46**(7), 1672–1681 (2011)
4. A. El Gamal, H. Eltoukhy, CMOS Image Sensors IEEE Circuits and Devices Magazine, vol. 21, no. 3 (2005), pp. 6–20
5. R. Sarpeshkar, T. Delbruck, C.A. Mead, White noise in MOS transistors and resistors. IEEE Circuits Devices Mag. **9**(6), 23–29 (1993)
6. O. Sgrott, D. Mosconi, M. Perenzoni, G. Pedretti, L. Gonzo, D. Stoppa, A 134-pixel CMOS sensor for combined Time-Of-Flight and optical triangulation 3-D imaging. IEEE J. Solid-State Circuits **45**(7), 1354–1364 (2010)
7. K. Oberhauser, G. Zach, H. Zimmermann, Active bridge-correlator circuit with integrated PIN photodiode for optical distance measurement applications, in Proceeding of the 5th IASTED International Conference Circuits, Signals and Systems, 2007, pp. 209–214
8. G. Zach, A. Nemecek, H. Zimmermann, Smart distance measurement line sensor with background light suppression and on-chip phase generation, in Proceeding of SPIE, Conference on Infrared Systems and Photoelectronic Technology III, vol. 7055, 2008, pp. 70550P1–70550P10
9. G. Zach, H. Zimmermann, A 2 × 32 Range-finding sensor array wit pixel-inherent suppression of ambient light up to 120klx, IEEE International Solid-State Circuits Conference (2009), pp. 352–353
10. G. Zach, M. Davidovic, H. Zimmermann, A 16 × 16 pixel distance sensor with in-pixel circuitry that tolerates 150 klx of ambient light. IEEE J. Solid-State Circuits **45**(7), 1345–1353 (2010)
11. P. Seitz, Quantum-noise limited distance resolution of optical range imaging techniques. IEEE Trans Circuits Syst **55**(8), 2368–2377 (2008)

12. A. Nemecek, K. Oberhauser, G. Zach, H. Zimmermann, Time-Of-Flight based pixel architecture with integrated double-cathode photodetector. in Proceeding IEEE Sensors Confcrcncc **278**, 275–278 (2006)
13. S. Decker, R.D. McGrath, K. Brehmer, C.G. Sodini, A 256 × 256 CMOS imaging array with wide dynamic range pixels and column-parallel digital output. IEEE J. Solid-State Circuits **33**(12), 2081–2091 (1998)
14. C. Bamji, H. Yalcin, X. Liu, E.T. Eroglu, Method and system to differentially enhance sensor dynamic range U.S. Patent 6,919,549, 19 July 2005

Sensors Based on In-Pixel Photo-Mixing Devices

Lucio Pancheri and David Stoppa

1 Introduction

The first Time-Of-Flight (TOF) 3D scanning systems were realized in the 1970s for military and space applications [1]. Both pulsed operation and CW modulation, either AM or FM, were used. Scanning systems have steadily improved in the following decades [2–4], evolving into commercial products for 3D metrology and modelling applications, although their cost remains high due to the requirements of the scanning mechanics.

The first technology enabling the realization of a scannerless range camera based on (TOF) principle is modulated image intensifier [5, 6], offering both high resolution and sub-mm precision at video rates. The cost of this technology, however, prevents this concept to be used in consumer applications.

A breakthrough in the field of 3D vision was observed with the introduction of devices capable of performing simultaneously light detection and demodulation. This class of detectors perform high frequency signal demodulation directly in the charge domain through semiconductor potential modulation, avoiding complex in-pixel mixing electronics. In this way, a very compact pixel layout can be obtained. The first successful charge demodulators were fabricated in CCD/CMOS technologies and presented in the 90s with the names of lock-in pixels or Photonic Mixing Devices (PMD) [7–9].

Although the first versions of demodulating devices were based on photogates, and thus required specialized CMOS/CCD technologies, other detector designs compatible with standard CMOS or CIS technologies have been presented in the last years, opening the way to the extension of the demodulation pixel principle to more advanced technology nodes [10–12]. In addition, two alternative

L. Pancheri (✉) · D. Stoppa
Fondazione Bruno Kessler, Via Sommarive 18 38123 Trento, Italy
e-mail: pancheri@fbk.eu

D. Stoppa
e-mail: stoppa@fbk.eu

F. Remondino and D. Stoppa (eds.), *TOF Range-Imaging Cameras*,
DOI: 10.1007/978-3-642-27523-4_4, © Springer-Verlag Berlin Heidelberg 2013

demodulation pixel concepts have successively been developed. The first one makes use of standard photodiodes coupled with switched-capacitor in-pixel electronics [13–15], while the second one exploits gated Single-Photon Avalanche Diodes [16, 17]. Although promising, the development of these last two approaches has not yet lead to commercial products. In fact, all the TOF 3D cameras currently on the market are based on charge-domain demodulation detectors approach [18].

The general principle of operation of monolithic TOF sensors based on charge-domain demodulation pixels will be illustrated in Sect. 2. The basic device structures starting from simple photogate demodulators to devices implemented in deep submicron and imaging technologies will be reviewed in Sect. 3, while pixel architectures will be covered in Sect. 4.

2 Basic Principle of Operation

In an Indirect TOF camera, each pixel independently performs homodyne demodulation of the received optical signal, and is therefore capable of measuring both its phase delay and amplitude. This kind of pixels are often referred to as demodulation pixels [19], correlating pixels [9] or lock-in pixels [7]. In this section, the general operation of a demodulation pixel will be illustrated, without entering the detail of the device physical structure.

An electro-optical demodulation pixel performs the following functions:

(a) Light detection
(b) Fast shutter or correlation
(c) Charge storage for multiple accumulations

An ideal correlating pixel can be modeled with a photo-current generator, a charge-flux control gate and a storage capacitor C_S, as shown in Fig. 1. If p(t) is the optical power incident on the pixel and g(t) is the transfer function of the control gate, the correlation function between p(t) and g(t) is defined by

$$c(\tau) = \lim_{T\to\infty} \frac{1}{T} \int_0^T p(t)g(t+\tau)dt, \tag{1}$$

where τ is the time delay between p(t) and g(t).

The output voltage on the storage capacitor after an integration time T_{int} is given by:

$$V(\tau) = \frac{R}{C_S} \int_0^{T_{int}} p(t)g(t+\tau)dt = \frac{R \cdot T_{int}}{C_S} c(\tau) \tag{2}$$

where R is the detector responsivity. The pixel output voltage $V(\tau)$ is therefore proportional to the correlation function $c(\tau)$ as defined in Eq. 1.

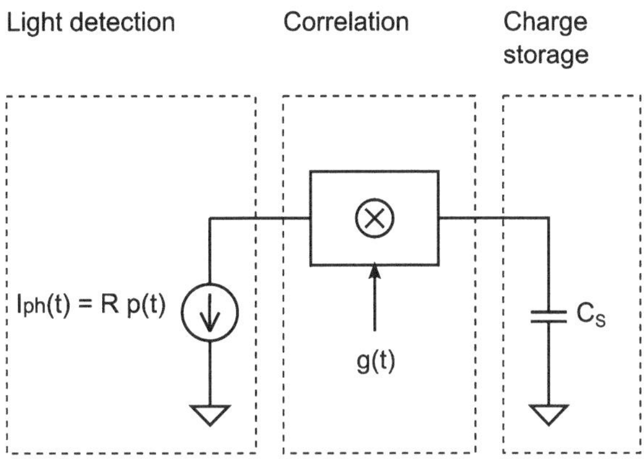

Fig. 1 Schematic illustration of a demodulation pixel

While in general g(t) can be a complex non-linear function of time, in most practical cases the control gate can be modeled as a fast shutter, and g(t) can be ideally described with a square wave switching between 0, when the charge is not integrated by the capacitor, and 1, when all the photo-charge is transferred to the capacitor. For a real device, the shutter efficiency is lower than 100 % and is a function of wavelength, frequency and angle of incidence of light on the detector [20]. Therefore, in general g(t) can range from a minimum value g_{min} to a maximum value g_{max}, and the non-ideality can be described with the detector demodulation contrast χ_D, defined as

$$\chi_D = \frac{g_{\max} - g_{\min}}{g_{\max} + g_{\min}}, \tag{3}$$

The optical power as a function of time p(t) can often be modeled as a sine-wave in a first approximation. The presence of higher-order harmonic components causes a measurement linearity distortion which needs to be corrected for an accurate result. The modulation depth χ_M of the light source, defined as the ratio between amplitude and offset, affects the measurement precision and must be taken into account. In fact, at the frequencies used in TOF cameras, which are in the tens of MHz range, it is difficult to modulate a high-power emitter with 100 % modulation depth.

The correlation function $c(\tau)$ between sine-wave optical signal p(t) and square-wave gating function g(t) is a sine wave function having amplitude A, offset B and a phase delay φ proportional to the time delay of the received optical signal, as shown in Fig. 2.

The overall demodulation contrast χ, defined as the ratio between amplitude A and offset B, measures the demodulation performance of the entire system including the illumination unit. In addition to the effects of detector, described by χ_D, and light source modulation depth χ_M, χ depends also on the ratio between sampling time and modulation period T. This effect, referred to as natural sampling

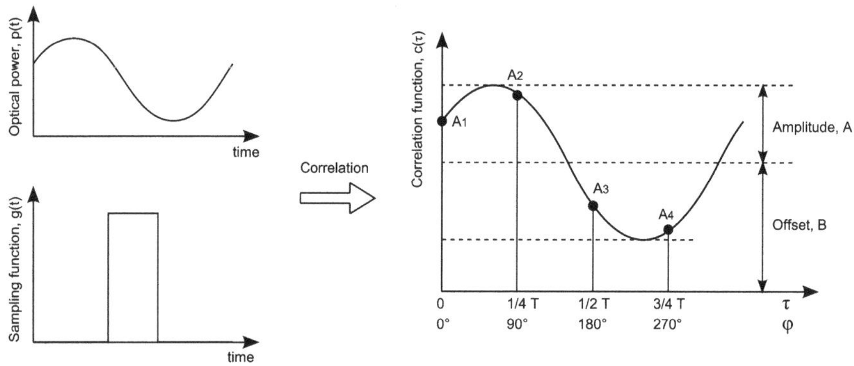

Fig. 2 Schematic illustration of the correlation function evaluation by 4-sample acquisition

[21], can be taken into account with an additional term χ_S. If the sampling time is much smaller than T, χ_S approaches 1 while with a sampling time of T/4 and T/2, which are more used in practice, χ_S is reduced to 0.9 and 0.64, respectively [21, 22]. The overall demodulation contrast χ of the system can be expressed as

$$\chi = (\chi_D \chi_S)\chi_M = \chi_P \chi_M. \tag{4}$$

The pixel demodulation contrast χ_P, which combines detector and sampling-related contrast terms, can be used as a figure of merit to quantify the pixel demodulation performance independently from the used light source.

The simplest approach to retrieve the phase delay of the optical signal consists in sampling the correlation function at four quadrant phase values (0°, 90°, 180° and 270°), as illustrated in Fig. 2. Using the four sampled values A_1-A_4 it is possible to calculate phase delay φ, amplitude A and offset B according to the following equations [23]:

$$\varphi = \arctan\left(\frac{A_1 - A_3}{A_2 - A_4}\right). \tag{5}$$

$$A = \frac{\sqrt{(A_1 - A_3)^2 + (A_2 - A_4)^2}}{2} \tag{6}$$

$$B = \frac{A_1 + A_2 + A_3 + A_4}{4} \tag{7}$$

A pixel with four shutters feeding four charge storage nodes allows the simultaneous acquisition of the four samples needed for the calculations in Eqs. 5, 6 and 7. The simplified schematic and the operation of a four-tap pixel are shown in Fig. 3. The light detection function is represented by a photo-current generator, where the current Iph is proportional to the incident optical signal p(t) through the responsivity R. The four shutters are activated one at a time for a time equal to T/4,

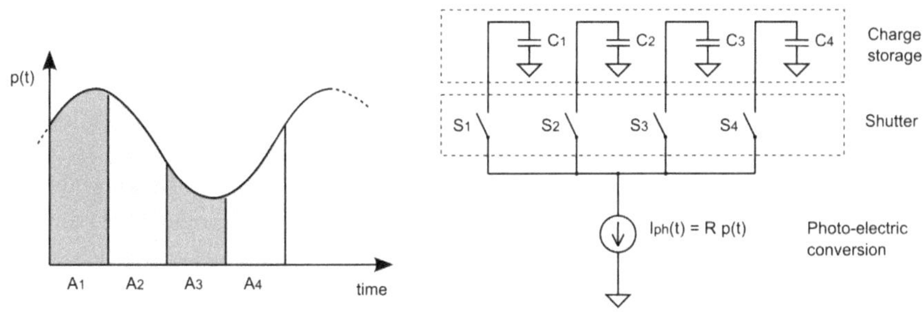

Fig. 3 Four-tap demodulation pixel operation principle and schematic diagram

and the shutter activation sequence is repeated for the whole integration time, which usually includes hundreds of thousands modulation periods.

A more commonly used structure in TOF cameras is the symmetric two-tap pixel, illustrated in Fig. 4. This pixel has only two shutters and two storage capacitors. Although with this configuration a compact pixel layout can be obtained, there are some drawbacks: in order to acquire the four samples required for phase calculation, two measurements have to be done: the first one to sample A_1 and A_3, and the second one for A_2 and A_4. Moreover, if background illumination is changing or the observed scene is moving during the acquisition, the measurement will be affected by motion artifacts.

An even simpler pixel can be obtained by removing one of the storage capacitance from the two-tap pixel and draining the corresponding charge out of the pixel. In this way, however, four measurements are necessary for phase calculation, thus reducing the overall frame rate and further increasing the problems of motion artifacts.

Although we have so far considered only the case of a sine-wave modulated illumination light, other solutions have been considered in the literature. Pulsed operation with low duty cycle can be conveniently employed to reduce the effect of ambient light [10]. Pseudo-noise modulation can be used to extend the non-ambiguous operation range [24] and in case of simultaneous operation of multiple cameras, to avoid interference between different systems [25].

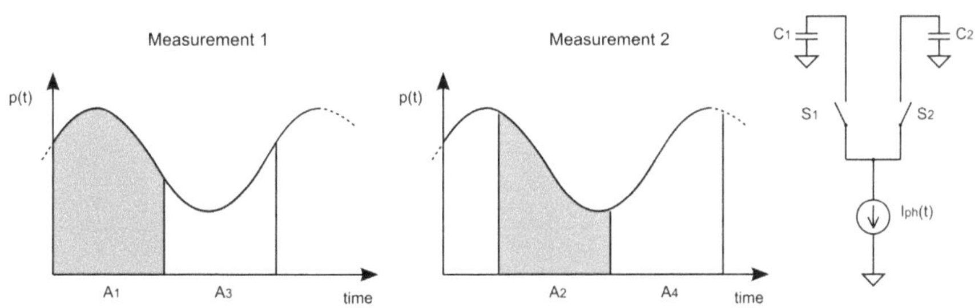

Fig. 4 Two-tap symmetric demodulation pixel operation principle and schematic diagram

While the most common transfer scheme can be represented with a two-valued transfer function g(t), in some cases g(t) can be sine-wave modulated, and the pixel is working as a sine-wave mixer [26]. This operation leads to a good measurement linearity, although the theoretical demodulation contrast that can be obtained in this case is smaller than for shutter operation.

In the next section, several types of demodulation pixels presented so far will be reviewed. Where possible, the main characteristics such as demodulation contrast, bandwidth, pixel size and fill factor, will be given, in order to give an idea of the quality of the device.

3 Charge-Domain Electro-Optical Demodulators

3.1 Basic Gate-Based Demodulators

The first demodulation pixels were developed in the 90s at PSI (Zürich, CH) [7], Carnegie Mellon University (Pittsburgh, US) [8] and University of Siegen (Siegen, DE) [9]. In all these works, the pixels were implemented in CMOS-CCD technologies, exploiting the fast charge transfer between CCD gates to obtain optical signal demodulation in the charge domain.

Although different gate topologies and pixel configurations have been proposed in subsequent works, two basic structures can be identified: namely the central-gate and the two-gate demodulator, which are shown in Fig. 5.

The central-gate demodulator consists of a photo-sensitive gate and two or more lateral transfer gates. In the earlier versions [7, 8] only the central gate is illuminated, while the lateral gates are covered with a metal shield. This structure

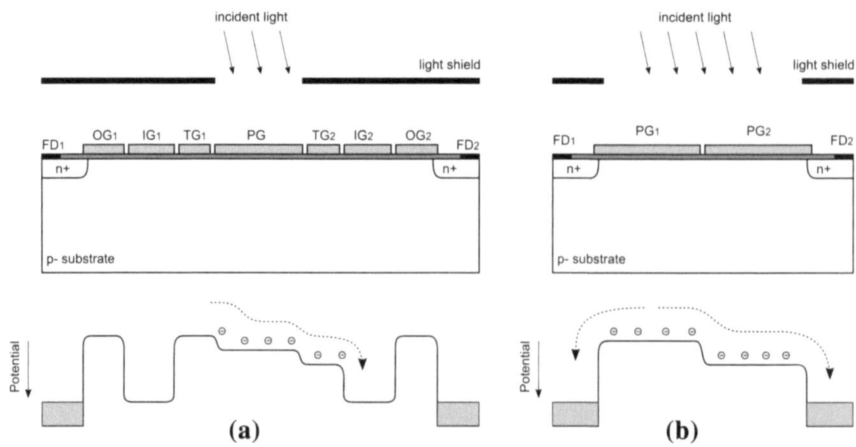

Fig. 5 Cross-section and potential profile of (**a**) central-gate demodulator with integration and transfer gates (**b**) two-gate demodulator

is shown in Fig. 5a, where two charge integration gates IG_1 and IG_2 and two output gates OG_1 and OG_2 are also present.

The potential profile is shown in Fig. 5a, when the voltage at transfer gate TG_1 is low and the voltage at TG_2 is high. Thanks to the potential energy gradient, electrons are collected by IG_2, while the electron flow can be directed towards IG_1 by inverting the transfer gates voltages.

The electrons at the border of the central gate are driven towards the integration nodes by drift due to the presence of a fringing field [20], while electrons far from the active transfer gate, where the lateral electric field is absent, move mainly by diffusion. The transfer time therefore increases with the size of the central gate: a small central electrode is thus needed in order to obtain a large demodulation bandwidth [20]. On the other hand, a smaller central gate determines a lower pixel fill factor and thus a lower sensitivity. A tradeoff between sensitivity and bandwidth is therefore present for this pixel.

Although the first demodulating pixels were fabricated using surface-channel CCD gates, successive designs have switched to buried channel CCD/CMOS processes. The presence of a buried channel enables a faster charge transfer due to the larger fringing field extension with respect to a surface-channel CCD [20].

Pixel design based on central gate demodulators having four transfer gates have also been proposed in [7, 27]. With this configuration, pixel-level sampling and storage of the correlation function at four quadrant phases is possible. The four-gate pixels presented so far feature a small fill factor, due to the area occupation of the additional gates and readout electronics. Moreover, before the phase delay is calculated using Eq. 5, a calibration needs to be performed to correct the gain non-uniformity due to geometrical mismatches [7].

Several modified versions of the central-gate pixel, fabricated in CMOS/CCD technologies, have been presented. A compact version featuring two transfer gates, but only an integration gate and an output gate was conveniently employed to realize one of the first TOF 3D camera prototypes [21]. In this pixel, one of the transfer gates is connected to the integration gate, while the other is used to transfer the charge to a draining diffusion. A 64 x 25 pixel sensor was implemented in a 2.0 μm CMOS/CCD process, with a 65 x 21 μm pixel size, a fill factor larger than 20 % and a 25 % demodulation contrast at 20 MHz in the near infrared. In this last implementation, the fill factor was enhanced by extending the illuminated region to the transfer gates, at the expense of a reduction of the detector demodulation contrast, which can be at most 67 % [21]. Moreover, the camera needs to acquire 4 successive 2D images to compute a 3D image.

The two-gate demodulator, known as Photonic Mixing Device (PMD) from the original works published in the late 90s [9], is a symmetric device with two equal polysilicon gates which are both illuminated and modulated. The photo-generated electrons move towards the two floating diffusions FD_1 and FD_2, according to the potential profile determined by gate voltages. A schematic cross section of the device is shown in Fig. 5b together with the potential profile. According to Fig. 5, in DC operation ¾ of the total photo-charge move towards one floating diffusion,

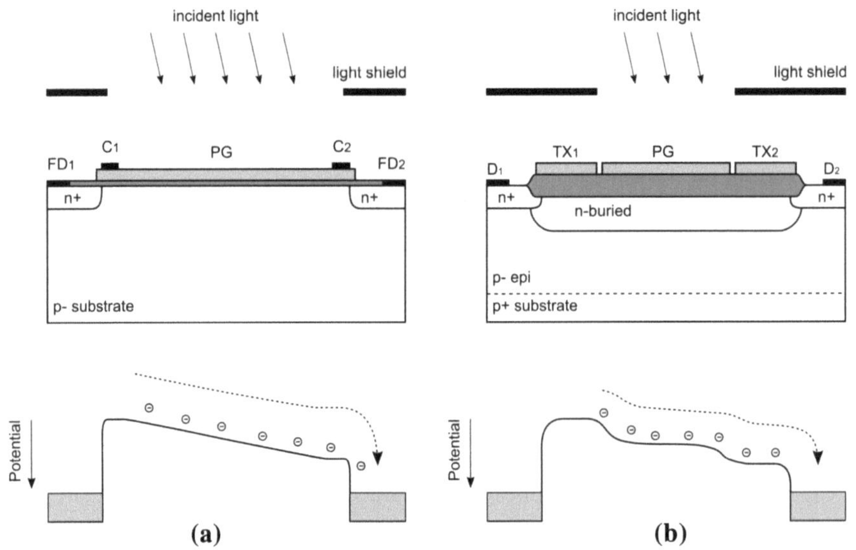

Fig. 6 Cross-section and potential profile of (**a**) high-resistivity gate demodulator (from [22]) (**b**) central-gate demodulator with gates on field oxide in 0.35 μm technology (from [10])

while the remaining ¼ move towards the other one. Thus, a maximum demodulation contrast of 50 % can be obtained with this device.

Commercial products based on this device have been on the market from several years [23]. An optimized design allowed obtaining demodulation contrast approaching 50 % at 100 MHz with 31 % fill factor in 40 x 40 μm pixels.

One of the requirements of gate-based demodulation pixels is obtaining a good demodulation contrast at high frequency. In order to obtain a good bandwidth, it is necessary to avoid the presence of any regions in the charge demodulation path where electrons move by diffusion. A demodulator structure addressing this issue is shown in Fig. 6a [22, 28]. A continuous high-resistivity gate is present on top of the device. The channel lateral potential varies continuously across the pixel, so that the electron transport is dominated by drift.

A four-tap version of this device fabricated in a surface-channel CCD/CMOS technology is described in [22]. The central polysilicon gate is a square with 25 x 25 μm area and the four modulation contacts and integration gates are placed at the edges. An average drift time of 3.4 ns was estimated for the implemented device, which could be theoretically reduced to 200 ps. Although DC shutter efficiency exceeds 70 % using infrared illumination, a demodulation contrast of only 26 % at 20 MHz was measured. This difference was ascribed to the slow diffusion of carriers from the substrate to the surface of the device, and could be improved by using a buried-channel CCD technology. The main disadvantages of this device are the static and dynamic power consumption related to the gate resistance and capacitance, which can cause local heating problems and require powerful driving electronics for large pixel arrays.

All the pixels described so far have been fabricated in dedicated CMOS/CCD technologies. Standard and imaging CMOS technologies do not provide CCD capabilities such as overlapping gates and buried channels. Therefore, scaling the device concepts developed in CCD technologies to advanced technology nodes require a process modification in the technology itself. However, in the last years several attempts have been done to implement demodulation pixels in standard and imaging CMOS technologies.

A CMOS version of the central-gate demodulator with photogates on top of field oxide was demonstrated in a 0.35 μm CMOS technology [10]. An additional processing step was used to create a buried channel below the field oxide. A cross section of the pixel is shown in Fig. 6b, where two draining electrodes, also present in the pixel, are not shown. A QVGA image sensor was fabricated, having a pixel size of 15 x 15 μm with 19 % fill factor. The sensor used 1 MHz modulation frequency with 10 % duty cycle, and could achieve a best-case range resolution of 2.35 cm.

Another CMOS-compatible photogate demodulator was demonstrated in a 0.18 μm CIS technology [11]. A cross section of the device is shown in Fig. 7 together with the potential profile along the cut-lines A-A'and B-B'. An n-type buried channel was obtained exploiting the tail of the buried diode nwell implantation through the polysilicon gates. A lightly doped surface p layer arising from the threshold adjustment implantation is present below the gate oxide. As can be observed in the potential profile plot, both a surface channel and a buried channel are present in this device. During the integration phase, most of the photo-generated electrons are collected in the buried channel, where a lateral electric field enables fast lateral charge transfer. The charge is then transferred from the buried channel to the surface channel of the gate with high bias voltage, and finally moves to the floating diffusion.

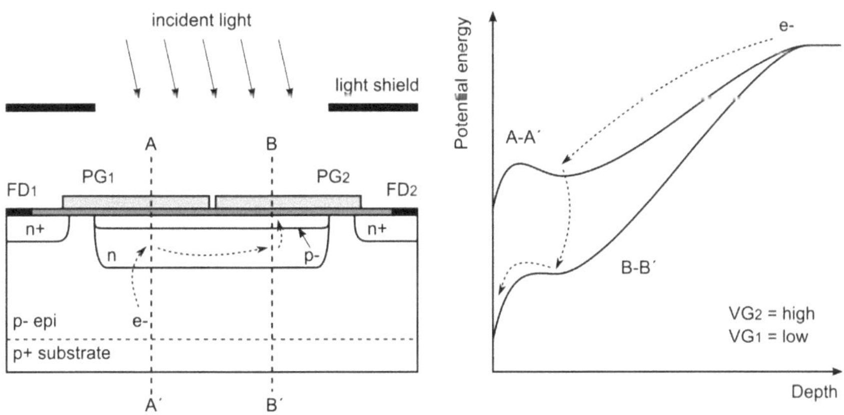

Fig. 7 Cross-section and, potential profile of buried channel demodulator in 0.18 μm CIS technology

A 60 x 80 pixel array based on this structure with 10-μm pixel pitch was fabricated in a 0.18 μm CMOS imaging technology. The pixel, having 24 % fill factor, could achieve a demodulation contrast of 40 % at 20 MHz and a bandwidth exceeding 50 MHz.

3.2 Current-Assisted Demodulators

While photogate-based pixels use gate voltage modulation to control the potential gradients at the semiconductor surface, an alternative approach exploits the electric field formed in the silicon substrate by modulating the voltage applied at two p+ substrate contacts [29]. Since a majority carrier current is associated to this field, the device exploiting this principle has been originally called Current Assisted Photonic Demodulator (CAPD).

A basic CAPD cross-section having two modulation electrodes $M_{1\text{-}2}$ and two floating diffusions $FD_{1\text{-}2}$ is shown in Fig. 8. The structure is compatible with standard CMOS technologies, provided that low-doped surface regions free from both n-wells and p-wells can be implemented. The electric field penetration depth into the substrate is proportional to the distance between the modulation electrodes. A good responsivity for near-infrared light wavelengths can thus be obtained with pixel sizes of a few tens of micrometers.

The main disadvantage of this pixel is the power consumption due to the majority current flow, which can be minimized using a high-resistivity silicon substrate [29, 30].

A 32 x 32 3D camera sensor based on this pixel was presented in a 0.35 μm CMOS technology [31]. The 30-μm pixels in this camera have a substrate contact along the border, while two floating diffusion surrounded by two ring-shaped

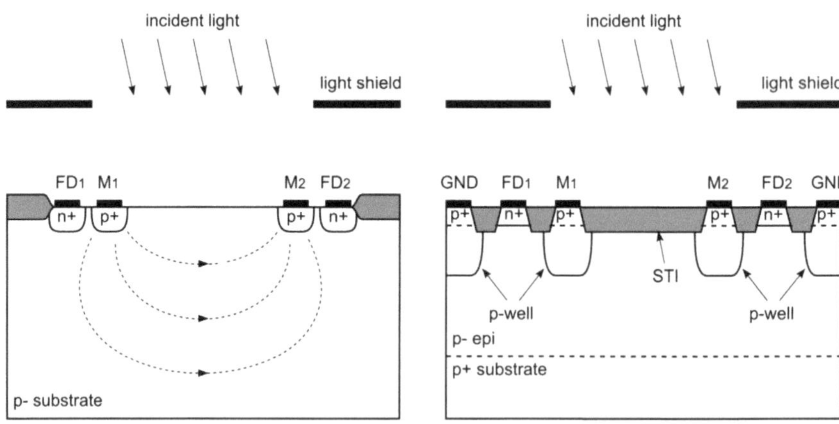

Fig. 8 Cross-section of Current Assisted Photonic Demodulator as originally presented in [29] and implemented in a 0.18 μm technology ([32])

modulation contacts are placed in the center of the pixel. The pixel features a 66 % fill-factor and reaches 51 % demodulation contrast at 20 MHz with a power consumption of 1.4mW per pixel, due to the modulation current.

A version of the CAPD demodulator realized in a 0.18 μm CMOS imaging technology is shown in Fig. 8b [32]. In the proposed device, fabricated in an epitaxial layer, two minimum-sized pwell diffusions are present under the modulation electrodes. A grounded pwell surrounding the device is used to host the n-type MOSFET readout electronics. Although the device was fabricated using a standard epitaxial layer doping, a demodulation contrast of 40 % at 20 MHz and a bandwidth larger than 45 MHz were obtained with a power consumption of only 10uW per pixel. A 120 x 160 pixel TOF sensor based on CAPD pixels was presented, having 10 μm pixel pitch and 24 % fill factor [33].

3.3 Pinned-Photodiode Demodulators

The idea of exploiting a pinned photodiode with multiple transfer gates for light demodulation was patented in the first 2000s [34]. While pinned photodiodes have been used in CMOS image sensors for more than a decade [35], one of the most demanding tasks in their design is the optimization of the transfer gate to obtain fast charge transfer. Residual potential barriers present between pinned diode and floating diffusion when the gate is on cause image lag [36] and make the device unsuitable for fast shutter operation. Moreover, the lack of a lateral electric field slows down the charge transfer through the gate and limits the maximum modulation frequency that can be achieved [37]. Several advancements have been done in the last years to improve the charge transfer speed of pinned photodiodes and make them suitable for TOF range imaging.

A symmetric pinned photodiode demodulator pixel with 10 μm pitch and 24 % fill factor fabricated in a 0.18 μm imaging technology was presented in [37]. A cross section of the device is shown in Fig. 9a. The device showed a demodulation contrast close to 100 % at DC, but its bandwidth was lower than 1 MHz.

A 12-μm pixel with 62 % fill factor was recently demonstrated using a 0.18 μm imaging process [37]. The pixel, using a single ended readout channel and a charge drain, features a demodulation contrast 35 % at 5 MHz.

The benefits of pixel scaling on device bandwidth are evident from another work presented by the same group [12], where the same pixel concept is ported to a 6-um pitch using a 0.11 μm imaging technology. In this work, a VGA RGB sensor capable of working as an IR range camera was presented. In range camera operation, 4 RGB pixels are binned to form a single range pixel thus obtaining a QVGA range image resolution. The presented pixels have 32.5 % fill factor and are operated at 10 MHz modulation frequency.

The optimization of pinned photodiodes for fast charge transfer applications has been recently studied by different research groups. Besides optical ranging, another application requiring fast shutter is time-resolved fluorescence spectroscopy.

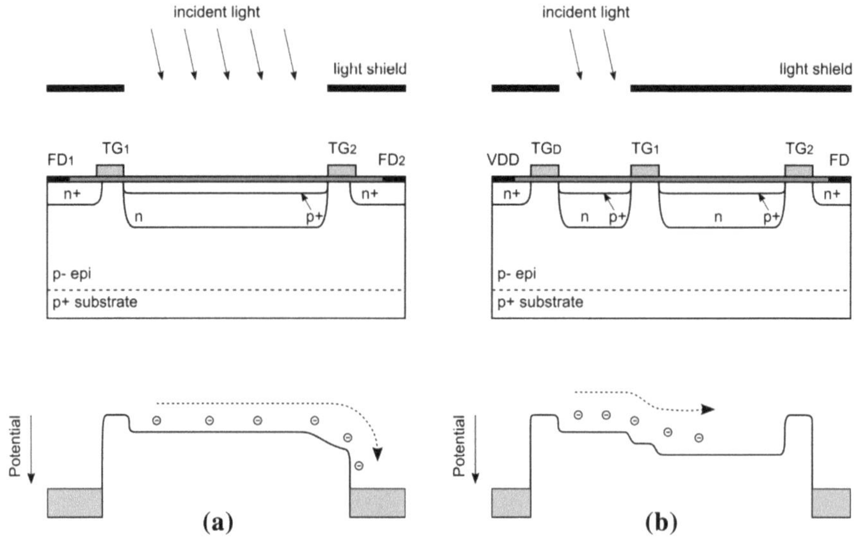

Fig. 9 Cross-section of (**a**) pinned photodiode demodulator [37] (**b**) two-stage pinned photodiode transfer pixel [39]

The dependence of the pinning potential on the diode lateral size was exploited to obtain a fast charge transfer from a small size pinned diode to a large size pinned diode [39]. A cross section of the pixel together with a potential profile is shown in Fig. 9. The use of another pinned diode for charge storage enables also the possibility to perform CDS operation, which would be lost if the charge were transferred directly to the floating diffusions. A drawback of this pixel is the reduced fill factor due to the small size of the illuminated diode.

The charge transfer speed in pinned photodiodes can be boosted by careful design of the transfer gate. In [40], the transfer time obtained using different transfer gate topologies is compared. It is experimentally demonstrated that the fastest transfer speed, with a complete charge transfer in less than 80 ns, is obtained with an U-shaped transfer gate, while using a linear transfer electrode similar to the one presented in [12], a complete charge transfer takes hundreds of ns. Thanks to the electrode U-shape, the residual potential barrier at the transfer gate is minimized, and a large charge demodulation bandwidth can be obtained.

3.4 Static Drift Field Demodulators

In the devices presented so far, electric field modulation affects the whole photosensitive pixel area, thus requiring modulation electrodes as large as the active area itself. Therefore, a high-power driving electronics needs to be implemented to

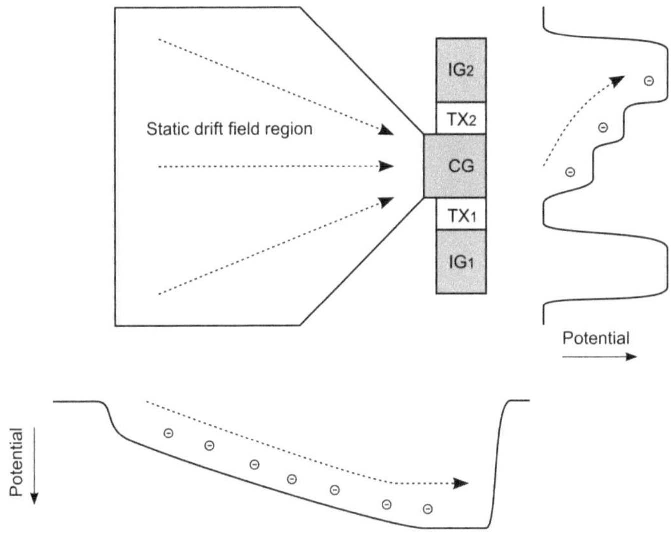

Fig. 10 Simplified layout and potential profile of Static Drift Field Demodulator

drive their large capacitance at high modulation frequencies. Moreover, in these pixels it is difficult to combine a high fill factor with a large demodulation bandwidth.

A powerful device concept addressing these issues is the static drift field demodulator [19], which is illustrated in Fig. 10. In this device, the large photo-sensitive area is physically separated from the small modulation region. In the former region, a static drift field is present, driving the photo-generated electrons towards the latter. The modulation region, implemented with a central-gate demodulator in Fig. 10, can be minimized to achieve a high bandwidth, high fill factor and small modulation capacitance.

The first implementation of a static drift field demodulator used a sequence of CCD gates biased at increasing voltage to generate the static drift field [19]. A simplified layout of the pixel is shown in Fig. 11a. The pixel, having 40 μm pitch, was implemented in a 0.6 μm CMOS/CCD technology, achieved 25 % fill factor and an estimated cutoff frequency of 105 MHz.

A static drift field pixel using a pinned photodiode was implemented by Tubert et al. [41]. Because of the dependence of pinning voltage on the diode size [39], a triangular shaped pinned photodiode has a built-in static drift filed, driving the photo-charges towards two transfer electrodes. The device, whose layout is shown in Fig. 11b, can achieve an estimated transit time of about 10 ns. A 128 × 128 test pixel array with a pitch of 11.6 μm was fabricated in a 90 nm CIS technology.

A similar approach using a different transfer gate geometry was demonstrated by Takeshita et al. [42]. Estimated charge transfer times are lower than 2 ns for triangle and horn-shaped pixels, compared to 500 ns for a rectangular diode of

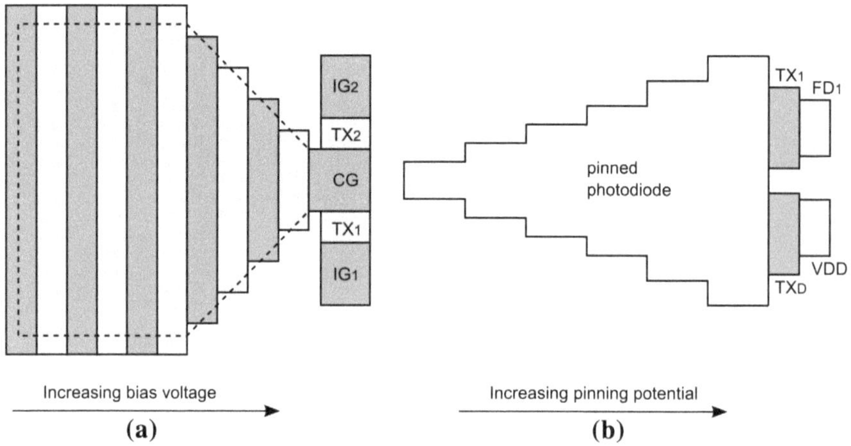

Fig. 11 Simplified layout of (**a**) photogate-based [19] and (**b**) triangle-shaped pinned photodiode [41] static drift field demodulators

similar size. Test pixels with 15 x 7.5 μm pixel size were implemented in a 0.18 μm CIS technology. Electro-optical tests performed with near IR LEDs have shown a demodulation bandwidth lower than expected due to the slow diffusion of photo-charges from the substrate.

Another approach that can be used to create a static drift field using pinned photodiodes is the creation of a pinning voltage gradient by means of a doping gradient in the n-well layer. Different methods have been proposed to create the required doping gradient. The most straightforward one is the use of three different n-type implantations, which however requires a non-standard modified process [43].

Another way to create a doping gradient in a pinned photodiode is using a striped mask to define the nwell implantation [44]. After annealing, the nwell stripes join in a unique nwell having a gradient in the doping profile, thus creating a lateral static drift filed. A 128 x 96 pixel range image sensor using this approach has been recently demonstrated in a modified 0.35 μm process [45]. Each pixel, having 40-μm pitch with 38 % fill factor, was based on a gradient-profile pinned photodiode feeding a central-gate demodulator with four transfer gates. The proposed pixel achieves a complete charge transfer from the pinned layer to the floating diffusion in less than 30 ns.

Engineering of pinning voltage gradients by combined geometry and doping effects has been demonstrated in a fast shutter pixel [46]. An additional p-type implantation has been used to reduce the potential of a pinned photodiode with respect to a pinned storage diode, while geometrical effects were used to create a linear potential gradient between the two. A draining gate along the channel enables an efficient and fast shutter mechanism with nanosecond resolution.

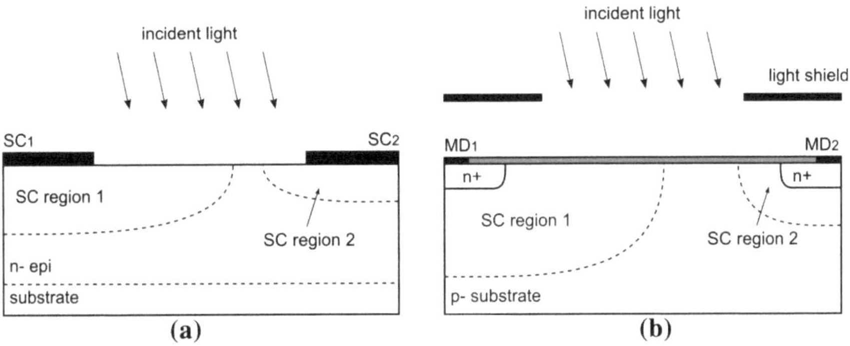

Fig. 12 Cross-section of (**a**) MSM demodulator (**b**) modulated space-charge region device

3.5 *Other Concepts*

Among the demodulating detectors presented so far for optical ranging applications there are Metal–Semiconductor-Metal (MSM) detectors, consisting of a pair of rectifying Metal–Semiconductor contacts (Schottky diodes) connected in series back-to-back [47]. In practice, MSM photosensors normally feature an interdigitated structure, that, in order to be operated properly, requires the applied voltage to be large enough to fully deplete the semiconductor surface region between the two electrodes. In this operating conditions, the capacitance between the two electrodes is very low, allowing a very fast operation of the detector. The cross section of a typical MSM device is shown in Fig. 12a.

In a MSM demodulator, the same electrodes are used to perform both voltage modulation and charge collection [24]. The photo-generated current enters from the positively-biased electrode (anode) and exits from the negatively-biased one (cathode). The photocurrent direction can be reversed by simply inverting the bias.

MSM devices operated as EOM in a range-finding system have been demonstrated in [26, 48]. Although this device is appealing for its excellent demodulation bandwidth and intrinsic background suppression characteristics, its non-compatibility with standard CMOS processes has prevented the fabrication of monolithic image sensors based on its principle.

Among less-explored demodulating device concepts there is the idea of using the space charge region modulation of two closely spaced junction diodes to direct the photo-generated electrons towards one or the other diode. A cross section of a device exploiting this concept is illustrated in Fig. 12b. This method has been described in a few works [24, 49–51], although a fully characterized range cameras has not been demonstrated so far. As in the case of the MSM device, the same electrodes are used for voltage modulation and charge collection. Therefore, it is not possible to use a standard active-pixel circuit, but it is necessary to implement in-pixel low-pass filtering stages to decouple the readout electronics from the modulation signal injected through decoupling capacitors.

A low-doped substrate or epitaxial layer is needed in order to obtain a good demodulation contrast from the variation of the space charge region widths. A metal shield must cover the n-type regions from direct illumination to reduce common-mode signal. In [24], the metal shield voltage is also modulated, to help the creation of a uniform potential gradient between the two electrodes.

4 Pixel Architectures

In most of the pixel concepts presented in this chapter, including photogate, current-assisted and pinned photodiode demodulators, the readout electrodes are different from the modulation electrodes. In this case, a simple 3T readout electronic channel can be implemented at the pixel level to perform charge-to-voltage conversion and buffering [10, 12, 21, 33].

Less explored demodulating detectors concepts, like MSM and modulated junction devices, require a more complex in-pixel readout electronics to decouple modulation and readout and to keep the electrode voltage at a constant common mode value [24, 50, 51]. In this case, area occupation and electronics power consumption limit the scalability of these pixel concepts and the implementation of high resolution arrays.

In photogate–based demodulators it is possible to perform Correlated Double Sampling (CDS) operation to reduce kTC noise thanks to the presence of an intermediate storage gate. This operation is not feasible if the demodulating device feed directly the floating diffusions. Pinned photodiode demodulators such as the ones presented in [12] use the transfer gate for charge demodulation and have therefore lost this possibility. However, CDS can be performed if a pinned storage diode with higher pinning voltage is implemented, as in [39].

Although most of the sensors use a sine-wave illumination, in some works a pulsed illumination with low duty cycle is used to improve ambient light immunity at the expense of sensor output linearity [10]. To exploit this possibility, it is necessary to implement a draining electrode on the pixel. Symmetric pixels can either be read-out differentially, thus achieving and enhanced dynamic range, or one of the two charge collection nodes can work as a charge drain, thus enabling the use of pulsed illumination. However, this second option can be usefully implemented only if the device has a shutter efficiency close to 100 %. As an alternative, a dedicated additional draining electrode can be implemented on-pixel, thus combining the benefits of multiple pixel outputs and low-duty cycle illumination [7, 10].

A fundamental characteristic affecting the sensor ultimate range resolution is the presence of an in-pixel offset subtraction circuit. Usually, only part of the sensor dynamic range is exploited to store the demodulated signal, while the remaining part is wasted by a common mode signal arising from dark current, ambient light and demodulation contrast lower than 100 %. An offset subtraction circuit allows to use all the available dynamic range to store the demodulated signal, and hence improves the sensor dynamic range, best range resolution and ambient light immunity.

A first approach for offset removal was proposed in [52], where an offset control gate was introduced. The control gate performs charge skimming operation on the integrated charge, subtracting a fixed amount of signal on each charge packet. Although an increased dynamic range is demonstrated, the amount of charge skimmed is fixed and not dynamically determined. The proposed implementation is therefore not useful in the case of strong variations either in the active signal or in background signal.

An evolution of this circuit is presented in [53], where a column-wise control circuit determines the amount of charge transferred to the sense nodes. In this implementation, the pixel has two integration gates, with a capacitance much larger than the one of the floating diffusion. At the end of the charge integration phase, the transfer gate voltage is increased slowly, allowing a gradual charge transfer from the integration gates to the floating diffusions. When a charge transfer from both integration gates is sensed, the transfer is stopped and the voltage difference is read out and digitized. This technique allows a factor 15 increase in the robustness against background light. The main disadvantages are an increased complexity of the pixel and column-level electronics and an decreased fill-factor.

Another technique employed to reduce the effect of background illumination is the injection of matched charge packets into the floating diffusions of a symmetric pixel. This method is schematically described in [22], although the details of the circuit are not disclosed. In this way, the difference signal is preserved, although the absolute value is destroyed. To be effective a pixel-level or column-level control circuit needs to be implemented to control the amount of injected charge.

Sensor dynamic range can be improved using pixel-wise integration. This method, described in [54], requires an in-pixel comparator and memory element, and is therefore suitable for large pixels only. This circuit solution needs to be implemented in a 4-tap pixel, so that an integration time common to all the taps is used.

5 Conclusion

In this chapter, the main demodulation pixel architectures presented so far have been reviewed, mainly focusing on the pixel core, the electro-optical demodulating detector. A transition can be observed from the first implementations in dedicated CMOS/CCD technologies to solutions using standard and CIS deep-submicron technology nodes. In particular, there is an effort to adapt the pinned photodiode, which is still not perfectly suited to the high frequency demodulation needs of TOF technique, by pushing its demodulation bandwidth through geometrical and technological modifications. Other promising approaches combining fast charge collection through static drift fields and high speed demodulation in a small mixing device are being experimented.

Each one of these approaches has his strengths and weaknesses, and there is not a clearly preferred embodiment suitable for all the potential applications. In the technology choice other factors should be considered in addition to quantum efficiency, fill factor, demodulation contrast, sensitivity and power consumption. The amount of non-standard processing steps to be implemented in a technology and the device portability and scalability for example are also important in perspective.

Even in the best commercial camera implementations, several aspects need to be improved. Ambient light immunity and dynamic range enhancement are two key features that can be handled at the device level, circuit level or system level, as well as the reduction of motion artifacts. As in conventional cameras, there is a demand of higher resolution and frame rate, as well as reduced power consumption. The addition of color in the same sensor is also a very appealing feature, and the first attempts to implement this functionality are being carried out.

The development of TOF technology should also take into account competing technologies, such as those based on pattern projection and stereo imaging. Although requiring heavy computation, they have already appeared on the market and have shown even more competitive than TOF in consumer applications, mainly because of their lower cost. The developments that TOF cameras will face in the next few years will surely show all the potential of this technology and will reveal if the evolution of TOF cameras in the 2010s will follow the same expansion experienced by conventional CMOS cameras in the 2000s.

References

1. P. Besl, Active optical range imaging sensors. Mach. Vis. Appl. **37**, 127–152 (1988)
2. K. Määttä, J. Kostamovaara, R. Myllylä, Profiling of hot surface by pulsed Time-Of-Flight laser range finder technique. Appl. Opt. **32**(27), 5334–5347 (1993)
3. P. Palojärvi, K. Määttä, J. Kostamovaara, Integrated Time-Of-Flight laser radar. IEEE Trans. Instrum. Meas. **46**(4), 996–999 (1997)
4. D. Dupuy, M. Lescure, and M. Cousineau, "A FMCW laser rangefinder based on a delay line technique" in Proc. IEEE Instrumentation and Measurement Technology Conference, pp. 1084–1088 (2001)
5. M. Kawakita, K. Iizuka, R. Iwama, K. Takizawa, H. Kikuchi, F. Sato, Gain-modulated Axi-Vision Camera (high speed high-accuracy depth-mapping camera). Opt. Express **12**, 5336–5344 (2004)
6. A.A. Dorrington, M.J. Cree, A.D. Payne, R.M. Conroy, D.A. Carnegie, Achieving sub-millimetre precision with a solid-state full-field heterodyning range imaging camera. Meas. Sci. Technol. **18**(9), 2809–2816 (2007)
7. T. Spirig, P. Seitz, O. Vietze, F. Heitger, The lock-in CCD two dimensional synchronous detection of light. IEEE J. Quantum Electron. **31**, 1705–1708 (1995)
8. R. Miyagawa, T. Kanade, CCD-based range-finding sensor. IEEE Transaction on Electron Devices **44**(10), 1648–1652 (1997)
9. R. Schwarte, Z. Xu, H. Heinol, J. Olk, R. Klein, B. Buxbaum, H. Fisher, J. Schulte, A new electrooptical mixing and correlating sensor: facilities and applications of the Photonic Mixer Device (PMD). Proc. SPIE **3100**, 245–253 (1997)

10. S. Kawahito, I.A. Halin, T. Ushinaga, T. Sawada, M. Homma, Y. Maeda, A CMOS Time-Of-Flight Range Image Sensor With Gates-on-Field-Oxide Structure. IEEE Sens. J. **7**(12), 1578–1586 (2007)
11. D. Stoppa, N. Massari, L. Pancheri, M. Malfatti, M. Perenzoni, L. Gonzo, A Range Image Sensor Based on 10-μm Lock-In Pixels in 0.18-μm CMOS Imaging Technology. IEEE J. Solid-State Circuits **46**(1), 248–258 (2011)
12. S.-J. Kim, J.D.K. Kim, S.-W. Han; B. Kang, K. Lee, C.-Y. Kim "A 640 × 480 image sensor with unified pixel architecture for 2D/3D imaging in 0.11 μm CMOS", IEEE Symposium on VLSI Circuits, pp. 92–93, June 2011
13. O. Elkhalili, O.M. Schrey, P. Mengel, M. Petermann, W. Brockherde, B.J. Hosticka, A 4 x 64 pixel CMOS image sensor for 3-D measurements applications. IEEE J. Solid-State Circuits **39**(7), 1208–1212 (2004)
14. D. Stoppa, L. Viarani, A. Simoni, L. Gonzo, M. Malfatti and G. Pedretti, "A 50x30-pixel CMOS Sensor for TOF-based Real Time 3D Imaging", *Proc. of the 2005 Workshop on Charge-Coupled Devices and Advanced Image Sensors*, (Nagano, Japan, 2005)
15. G. Zach, M. Davidovic, H. Zimmermann, A 16 x 16 Pixel Distance Sensor With In-Pixel Circuitry That Tolerates 150 klx of Ambient Light. IEEE J. Solid-State Circuits **45**(7), 1345–1353 (2010)
16. C. Niclass, C. Favi, T. Kluter, F. Monnier, E. Charbon, Single-Photon Synchronous Detection. IEEE J. Solid-State Circuits **44**(7), 1977–1989 (2009)
17. L. Pancheri, N. Massari, F. Borghetti, D. Stoppa, *A 32x32 SPAD Pixel Array with Nanosecond Gating and Analog Readout* (Proc, IISW, 2011). (Hokkaido, Japan, 2011)
18. See for example: PMD technologies (www.pmdtec.com), MESA imaging AG (www.mesa-imaging.ch), SoftKinetic (www.softkinetic.com), Panasonic (www.pewa.panasonic.com), Fotonic (www.fotonic.com)
19. B. Buttgen, F. Lustenberger, P. Seitz, Demodulation Pixel Based on Static Drift Fields. IEEE Trans. Electron Devices **53**(11), 2741–2747 (2006)
20. R. Lange, *3D Time-Of-Flight distance measurement with custom solid-state image sensors in CMOS/CCD-technology* (University of Siegen, Siegen, Germany, PhD dissertation, 2003)
21. R. Lange, P. Seitz, Solid-state Time-Of-Flight range camera. IEEE J. Quantum Electron. **37**(3), 390–397 (2001)
22. B. Büttgen, T. Oggier, R. Kaufmann, P. Seitz, N. Blanc, Demonstration of a Novel Drift Field Pixel Structure for the Demodulation of Modulated Light Waves with Application in Three-Dimensional Image Capture. Proc. SPIE **5302**, 9–20 (2004)
23. T. Möller, H. Kraft, J. Frey, M. Albrecht, R. Lange, *"Robust 3D Measurement with PMD Sensors", Proceedings of the 1st Range Imaging Research Day at ETH* (Zurich, Switzerland, 2005)
24. R. Schwartc, Dynamic 3D vision. Proceedings EDMO **2001**, 241–248 (2001)
25. B. Büttgen, M.-A. El Mechat, F. Lustenberger, P. Seitz, Pseudonoise Optical Modulation for Real-Time 3-D Imaging With Minimum Interference. IEEE Transactions on Circuits and Systems I **54**(10), 2109–2119 (2007)
26. P. Gulden, D. Becker, M. Vossiek, Novel optical distance sensor based on MSM technology. IEEE Sensors J. **4**(5), 612–618 (2004)
27. R. Lange, P. Seitz, A. Biber, S. Lauxtermann, Demodulation Pixels in CCD and CMOS Technologies for Time-Of-Flight Ranging. Proc. SPIE **3965**, 177–188 (2000)
28. P. Seitz, "Image sensing device and method of", US patent 2006/0108611A1
29. D. Van Nieuwenhove, W. Van Der Tempel, M. Kuijk, *Novel standard CMOS detector using majority current for guiding photo-generated electrons towards detecting junctions* (Proc. IEEE/LEOS Symp, Benelux Chapter, 2005), pp. 229–232
30. L. Pancheri, D. Stoppa, N. Massari, M. Malfatti, C. Piemonte, G.-F. Dalla Betta,"Current Assisted Photonic Mixing Devices fabricated on High Resistivity Silicon", Proc. IEEE Sensors 2008, (Lecce,Italy, 2008), pp. 981–983
31. W. van der Tempel, R. Grootjans, D. Van Nieuwenhove and M. Kuijk "A 1 k-pixel 3-D CMOS sensor", *Proc. IEEE Sensors Conference*, pp.1000–1003 (2008)

32. G.-F. Dalla Betta, S. Donati, Q.D. Hossain, G. Martini, L. Pancheri, D. Saguatti, D. Stoppa, G. Verzellesi, Design and Characterization of Current-Assisted Photonic Demodulators in 0.18-μm CMOS Technology. IEEE Trans. Electron Devices **58**(6), 1702–1709 (2011)
33. L. Pancheri, D. Stoppa, N. Massari, M. Malfatti, L. Gonzo, Q. D. Hossain, G.-F. Dalla Betta, "A 120x160 pixel CMOS range image sensor based on current assisted photonic demodulators", *Proc. SPIE*, Vol. 7726, 2010, pp. 772615 (SPIE Photonics Europe, Brussels, Belgium, 2010)
34. V. Berezin, A. Krymski, and E. R. Fossum, "Lock-in pinned photodiode photodetector," US Patent 2003/0213984A1, Nov. 2003
35. A. Theuwissen, *CMOS Image Sensors: State-Of-The-Art and Future Perspectives* (Proc, IEEE ESSDERC, 2007). (Munich, Germany, 2007)
36. E.R. Fossum, "Charge Transfer Noise and Lag in CMOS Active Pixel Sensors", *Proc. of 2003 IEEE Workshop on Charge-Coupled Devices and Advanced Image Sensors* (Schloss Elmau, Bavaria, Germany, May 2003)
37. D. Stoppa, N. Massari, L. Pancheri, M. Malfatti, M. Perenzoni, and L. Gonzo, "An 80 x 60 Range Image Sensor based on 10um 50 MHz Lock-In Pixels in 0.18um CMOS", *Proc. ISSCC 2010*, San Francisco, 7-11 Feb. 2010
38. S.-J. Kim, S.-W. Han, B. Kang, K. Lee, J.D.K. Kim, C.-Y. Kim, A Three-Dimensional Time-Of-Flight CMOS Image Sensor With Pinned-Photodiode Pixel Structure. IEEE Electron Device Lett. **31**(11), 1272–1274 (2010)
39. H.-J. Yoon, S. Itoh and S. Kawahito "A CMOS image sensor with in-pixel two-stage charge transfer for fluorescence lifetime imaging", *IEEE Trans. Electron Devices*, Vol. 56, No. 2, pp.214–221 (2009)
40. L.-E. Bonjour, T. Baechler, M. Kayal, *High-Speed General Purpose demodulation pixels based on buried photodiodes* (Proc, IISW, 2011). (Hokkaido, Japan, 8.–11. June 2011)
41. C. Tubert, L. Simony, F. Roy, A. Tournier, L. Pinzelli, P. Magnan, *High Speed Dual Port Pinned-photodiode for Time-Of-Flight Imaging* (Proc, IISW, 2009). (Bergen, Norway, June 26.-28. 2009)
42. H. Takeshita, T. Sawada, T. Iida, K. Yasutomi, S. Kawahito, High-speed charge transfer pinned-photodiode for a CMOS Time-Of-Flight range image sensor. Proc. SPIE **7536**, 75360R (2010)
43. S.J. Kim, S.W. Han, "Image sensor and operating method", US patent 2011/0198481 A1
44. D. Durini, A. Spickermann, R. Mahdi, W. Brockherde, H. Vogt, A. Grabmaier, B.J. Hosticka, Lateral drift-field photodiode for low noise, high-speed, large photoactive-area CMOS imaging applications. Nuclear Instruments and Methods in Physics Research A **624**(2), 470–475 (2010)
45. A. Spickermann, D. Durini, A. Suss, W. Ulfig, W. Brockherde, B.J. Hosticka, S. Schwope, A. Grabmaier, CMOS 3D image sensor based on pulse modulated Time-Of-Flight principle and intrinsic lateral drift-field photodiode pixels. Proc. ESSCIRC **2011**, 111–114 (2011)
46. S. Kawahito, Z. Li, K. Yasutomi, "A CMOS image sensor with draining only demodulation pixels for time-resolved imaging," *Proc. IISW 2011*, pp.185–188, (Hokkaido, Japan, 8.–11. June 2011)
47. S.M. Sze, D.J. Coleman Jr, A. Loya, Current transport in metal-semiconductor- metal (MSM) structures. Solid-State Electon. **14**, 1209–1218 (1971)
48. H. Kraft, J. Frey, T. Moeller, M. Albrecht, M. Grothof, B. Schink, H. Hess, B. Buxmbaum, *"3D-camera of high 3D-frame rate, depth-resolution and background light elimination based on improved PMD (Photonic Mixer Device)-technologies", Proc, OPTO 2004* (Nuernberg, Germany, 2004)
49. C. Bamji, H. Yalcin, "Methods and devices for improved charge management for three-dimensional and color sensing", US patent 7,352,454 B2
50. F. De Nisi, D. Stoppa, M. Scandiuzzo, L. Gonzo, L. Pancheri, G.-F. Dalla Betta, "Design of electro-optical demodulating pixel in CMOS technology", *Proceedings of IEEE International Symposium on Circuits and Systems (ISCAS 2005)*, pp. 572–575, (Kobe, Japan, 23–26 May 2005)

51. K. Oberhauser, G. Zach, A. Nemecek, H. Zimmermann, *Monolythically Integrated Optical Distance Measurement Sensor with Double-Cathode Photodetector* (Proc, IMTC, 2007). (Warsaw, Poland, May 1–3, 2007)
52. T. Spirig, M. Marley, P. Seitz, The multitap lock-in CCD with offset subtraction. IEEE Trans. Electron Devices **44**(10), 1643–1647 (1997)
53. T. Oggier, R. Kaufmann, M. Lehmann, B. Büttgen, S. Neukom, M. Richter, M. Schweizer, P. Metzler, F. Lustenberger, N. Blanc, Novel Pixel Architecture with Inherent Background Suppression for 3D Time-Of-Flight Imaging. Proc. SPIE **5665**, 1–8 (2005)
54. M. Lehmann, T. Oggier, B. Büttgen, C. Gimkiewicz, M. Schweizer, R. Kaufmann, F. Lustenberger, N. Blanc, *Smart Pixels for Future 3D-TOF Sensors* (Proc, IISW, 2005). (Nagano, Japan, 9.-11. June 2005)

Understanding and Ameliorating Mixed Pixels and Multipath Interference in AMCW Lidar

John P. Godbaz, Adrian A. Dorrington and Michael J. Cree

1 Introduction

With the advent of cheap full-field amplitude modulated continuous wave (AMCW) lidar systems range-imaging has become a technology with the potential for widespread applications in fields such as gaming, human-computer interface design, process-line quality control and vehicular warning systems. While off-the-shelf commercial systems offer centimetre level precision, the accuracy of measurements is frequently an order of magnitude worse due to uncalibrated systematic errors, resulting primarily from mixed pixels and multipath interference. In order to fully utilise the range-images produced by AMCW lidar range-cameras it is important to understand the limitations of the measurement process and the most common approaches to mitigation of these errors.

AMCW lidar systems are fundamentally limited in that each pixel in a full-field system, or point-sample for a point scanning system, is capable of measuring the range to only a single object. If there are multiple returns, due to sampling near the edge of an object or due to crosstalk between measurements, then erroneous range and amplitude data are produced. This is the mixed pixel/multipath interference problem, and forms the subject of this chapter.

In the remainder of Sect. 1.1 we explain the AMCW measurement technique in the context of sampling the spatial frequencies of backscattered signal returns within each pixel and the conditions for the occurrence of the systematic errors caused by multiple returns within a pixel. By developing a detailed model of measurement formation in Sect. 2, we explain the nature of the perturbations introduced by mixed pixels and multipath interference. In Sects. 3 and 4, starting from the first reports of mixed pixels in point-scanning AMCW systems, we cover

J. P. Godbaz (✉) · A. A. Dorrington · M. J. Cree
School of Engineering, University of Waikato, Hamilton, New Zealand
e-mail: jpg7@waikato.ac.nz

F. Remondino and D. Stoppa (eds.), *TOF Range-Imaging Cameras*,
DOI: 10.1007/978-3-642-27523-4_5,

the gamut of research over the past two decades into mixed pixels and multipath interference, including a variety of detection and mitigation techniques that can be applied to range-data from standard commercial cameras.

1.1 Time of Flight Measurement Techniques

All lidar systems operate using the Time-Of-Flight (TOF) principle. Because light travels at a known finite speed, it is possible to determine the range to an object using an active illumination source and measuring the range-dependent propagation delay induced in the illumination signal as it travel to and from objects in the scene.

A basic overview of the TOF principle is illustrated in Fig. 1. An illumination source sends out pulses of light; for the light to be measured by a sensor co-located with the illumination source it must travel twice the distance to the target object (the stick figure in Fig. 1). The backscattered signal intensity, as a function of range, can be represented by the backscattering function $f_\xi(r)$, as shown in Fig. 2a for the case of a single backscattering source such as the stick figure. Lidar systems operate by indirectly sampling this function in different ways depending on the type of lidar system. If another object is placed half way between the target and the illumination source, then the light from the nearer object is reflected back to the camera and arrives before light from the stick figure; $f_\xi(r)$ for this case is shown in Fig. 2b. This is the multiple return case; depending on the modulation technique, not all systems can correctly interpret this situation.

Fig. 1 The Time-Of-Flight principle. **a** Illuminating the scene. **b** Measuring the Time-Of-Flight to determine range

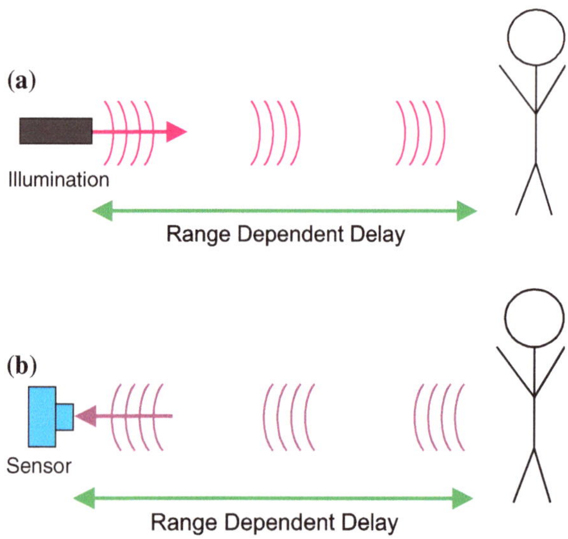

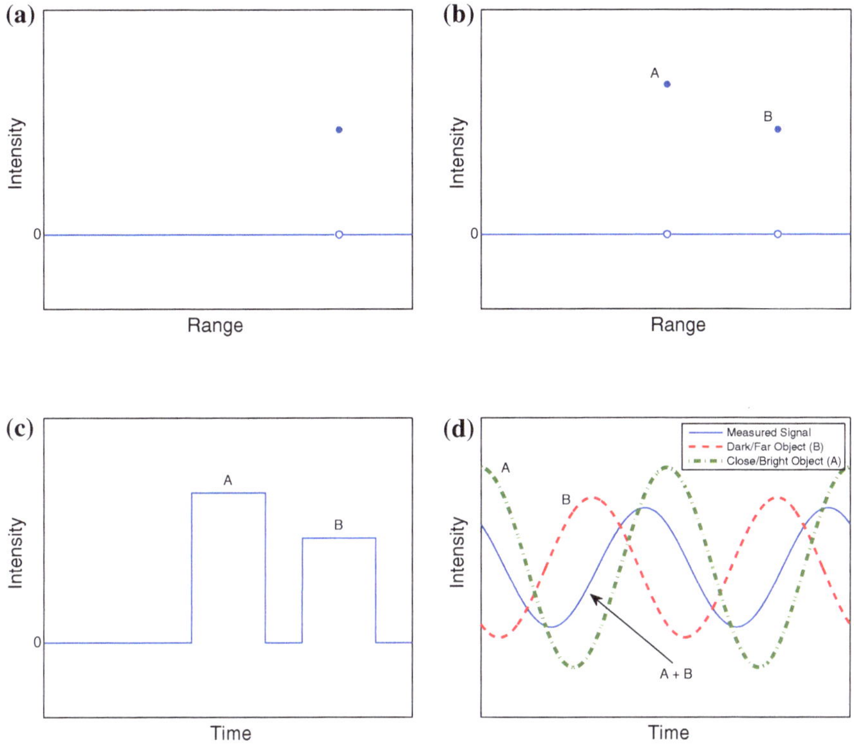

Fig. 2 The backscattered illumination waveforms for different modulation techniques in the two component case. **a** A single backscattering source. **b** Two backscattering sources. **c** Measured illumination (range-gating, two returns). **d** Measured illumination (AMCW, two returns)

In order to understand mixed pixels and multipath interference it is necessary to first understand how lidar systems sample the backscattering function $f_\xi(r)$. Examples of three different types of modulation are plotted in Fig. 3: the first is amplitude modulated continuous wave (AMCW, Fig. 3a), the second is frequency modulated continuous wave (FMCW, Fig. 3c) and the last is range-gating or pulsed modulation (Fig. 3e). As the illumination modulation, Ψ_i, travels from the camera to the objects in the scene and back, reflections from different objects are superimposed. As a result, the illumination signal over time at the sensor is given by,

$$\Psi_m(t) = f_\xi * \Psi_i, \tag{1}$$

or in the Fourier domain

$$\Psi_m(u) = F_\xi(u)\Psi_i(u), \tag{2}$$

where '$*$' represents convolution. Depending on the spectral content of the illumination modulation waveform over time, one can make inferences about the nature of the backscattering function. Only the spatial frequencies of the backscattering

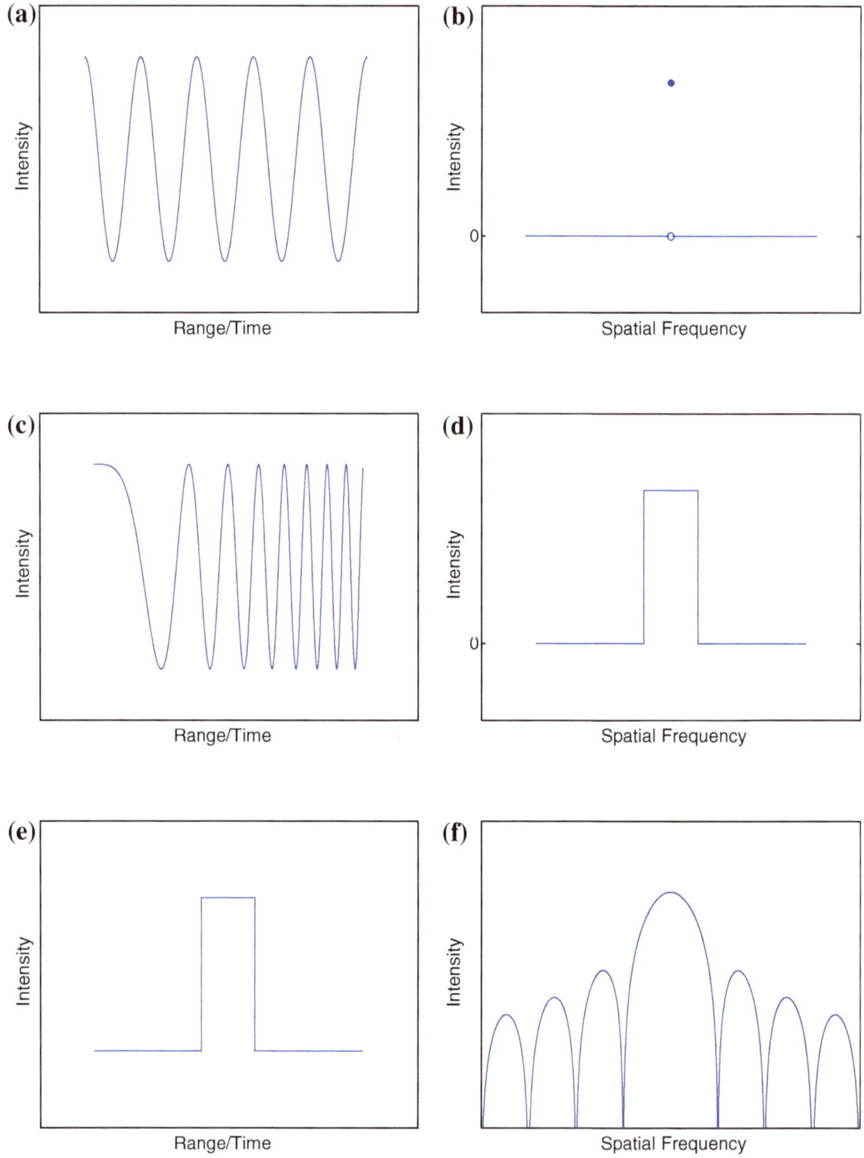

Fig. 3 A comparison of the illumination waveforms and the spatial frequencies they implicitly sample for different lidar modulation techniques. **a** AMCW illumination waveform. **b** AMCW spatial frequencies. **c** FMCW illumination waveform. **d** FMCW spatial frequencies. **e** Range-gating illumination waveform. **f** Range-gating spatial frequencies

function that are also present in the illumination modulation waveform are capable of being measured by the sensor. Therefore if the illumination modulation is perfectly sinusoidal, as in ideal AMCW, it is only possible to reconstruct a single spatial frequency of the signal returns. While this is enough to recover the range and amplitude to a single backscattering source, it is not enough information to separate out two superimposed returns.

Figure 2c, d show the backscattered illumination modulation, as measured at the sensor, for the range-gating and AMCW techniques given the backscattering function of Fig. 2b. Whereas the two returns are clearly separated in the range gating case, in the AMCW case the superposition of two sinusoids results in another sinusoid, which is indistinguishable from that produced by a single return (see Fig. 2d). When the phase and amplitude of the illumination are sampled, the estimated amplitude and range will be erroneous and correspond to neither of the backscattering sources.

Even though lidar modulation techniques are exactly the same as those utilised in radar, no practical radar systems appear to use AMCW: one of the primary reasons is the issue of multipath interference. Multipath interference and mixed pixels result when a single range measurement contains light from more than one backscattering source; in an AMCW system this leads to erroneous range and amplitude measurements. The details of formation are explained in Sect. 1.2 below. The radar community has moved away from ACMW to more sophisticated techniques, and as a result little has been published on its intricacies outside of the lidar literature. For full-field systems, however, AMCW offers advantages over FMCW and range-gating systems because it requires much lower data rates. This is particularly crucial to producing full-field range images at a high frame rate.

1.2 The Formation of Mixed Pixels and Multipath Interference

Errors due to mixed pixels and multipath interference occur in AMCW systems, but rarely in FMCW systems or range-gating systems, because AMCW systems only sample a single discrete spatial frequency. The terms *mixed pixel* and *multipath interference* distinguish different mechanisms that lead to the same end result. Mixed pixels occur because of the imaging nature of full-field systems or due to the finite size of the illumination spot in point-scanning systems. In full-field systems, if the image formed on the senor is out of focus, two objects can blur together onto one pixel, hence multiple AMCW returns are combined in that pixel. This is a reasonably common situation because the quality of the range estimate is generally linked to the received optical power, thus there is a temptation to use large apertures in order to collect more light, thereby degrading depth-of-field. Even if the image is in perfect focus, mixed pixels can still occur at object edges, where both the foreground and background objects are integrated by a single pixel.

Figure 4 is a range-image of a scene suffering from both mixed pixels and multipath interference. Region A is the edge of an object outside of the depth-of-field of the current focal parameters, thus resulting in defocus induced mixed pixels.

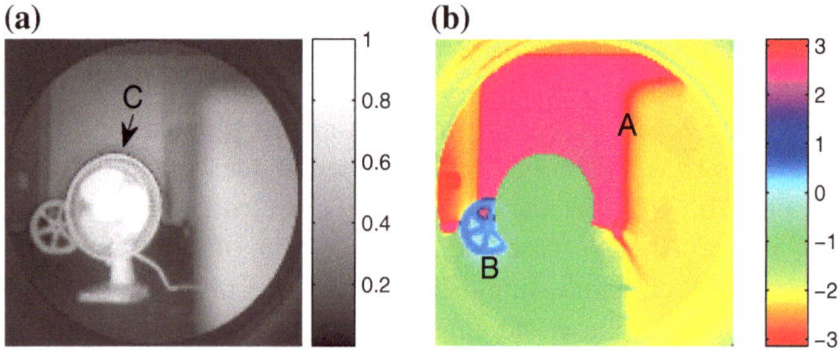

Fig. 4 An example scene, captured using a full-field AMCW range-imager, suffering from both mixed pixels (*A*, *C*) and multipath interference (*B*). **a** Amplitude. **b** Phase

Region C also suffers from mixed pixels; in this case the fan and the background are almost 180° out of phase, thus the mixed pixels result in amplitude cancellation. Common causes of mixed pixel and multipath interference are shown in Fig. 5, and include integration over an object edge and integration over a region subject to defocus blurring. It is also possible to get mixed pixels from thin structures that are thinner than the width of a pixel. In some cases transverse motion may be deliberately encoded as mixed pixels by using multiple accumulators on the sensor; by alternating between accumulation of each phase step it is possible convert the

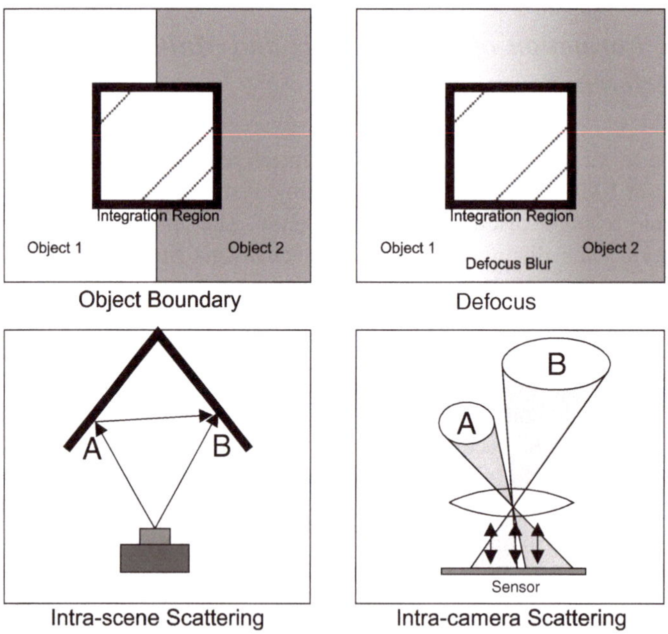

Fig. 5 Common causes of mixed pixels (*top row*) and multipath interference (*bottom row*)

motion problem to a mixed pixel restoration problem. This is particularly suited to multifrequency methods (see Sect. 4.6).

Multipath interference refers to the situation where light from one part of the scene is scattered onto pixels imaging a different part of the scene, causing interference. This occurs in full-field systems due to the simultaneous illumination of the entire scene. The interfering light can originate from a variety of sources/mechanisms and can be easily observed in high contrast scenes where bright objects cause erroneous measurements in darker objects. Although multipath interference still occurs in lower contrast scenes, it is not as obvious. Intra-camera lens scattering is one common cause, where light impinging on one region of the sensor is not completely absorbed and is reflected back towards the lens. The surface of the lens then reflects and scatters a portion of this light back onto other areas of the sensor, causing interference.

Multipath interference can also be caused by scattering and reflections in the scene. Ideally, objects in the scene are illuminated directly and solely from the illumination source, but scattering can occur from bright or reflective objects, causing additional interfering illumination with larger path lengths, perturbing the measurement data. Region B in Fig. 4 corresponds to multipath interference induced by highly localised scattering within the optics of the imaging system; generally intra-camera scattering is relatively homogeneous but the particular system [1, 2] used to capture Fig. 4 has non-standard characteristics due to an image-intensifier based design. The two most common causes of multipath interference are shown in the bottom row of Fig. 5. In the intra-scene scattering case, measurements of region B are perturbed due to illumination scattered onto region B by region A. In the intra-camera scattering case, regions A and B are imaged onto different parts of the sensor. In an ideal situation they would be completely independent measurements, however, in practice light scattering results in crosstalk between the measurements of A and B.

Although we concentrate on signal/image processing approaches to handling mixed pixels and multipath interference, it is also possible to ameliorate these effects using hardware. For example, by using anti-reflection coatings on lenses and by matching the angular width of the illumination to the camera field-of-view, to avoid perturbations from outside the field-of-view. Theoretically, one could decrease the incidence of mixed pixels by decreasing the fill-factor/sensitive region of each pixel in a full-field system, attempting to approach the ideal of sampling over an infinitesimal solid angle, but this would reduce SNR unacceptably.

2 Modelling Mixed Pixels and Multipath Interference

In this section we develop a detailed model of measurement formation, basing our notation on the model of Godbaz et al. [3] and continuing from the description in Sect. 1.

2.1 The Formation of Range Measurements

Each pixel on the sensor integrates over a particular solid angle of the scene: the intensity of the backscattered signal within an arbitrary pixel as a function of range is notated as $f_\xi(r)$, where r is the radial distance from the camera. The standard operating assumption is that there is a single component return within the pixel, namely

$$f_\xi(r) = a_0\delta(r - r_0) \tag{3}$$

$$F_\xi(u) = a_0 e^{-2\pi j u r_0}, \tag{4}$$

where $F_\xi(u)$ is the Fourier transform of the backscattered signal return intensity versus range, j is the imaginary unit, and where a_0, is the amplitude and r_0 is the range to the backscattering source.

AMCW lidar systems operate by indirectly measuring the backscattered illumination signal. In CMOS sensor designs, the backscattered illumination is correlated at the sensor with a sensor modulation waveform. The measured intensity at the sensor as a function of the relative phase, ϕ, of the illumination and sensor modulation signals can be written as

$$h(\phi) = \int_{-\pi}^{\pi}\int_{-\pi}^{\pi} f_\xi\left(\frac{\lambda}{4\pi}(\theta' + \phi')\right)\Psi_i(-\theta')\Psi_s(\phi' + \phi)\, d\theta' d\phi' \tag{5}$$

$$H(u) = F_\xi\left(\frac{4\pi}{\lambda}u\right)\Psi_i(u)\Psi_s^*(u), \tag{6}$$

where $\Psi_i(\phi)$ is the illumination modulation waveform as a function of phase, $\Psi_s(\phi)$ is the sensor modulation waveform and λ is the modulation wavelength. The waveform $h(\phi)$ is often referred to as the correlation waveform and is the device through which AMCW lidar measurements are achieved.

In the ideal case of a perfectly sampled correlation waveform with a modulation wavelength of λ, a complex domain range measurement is given by

$$\xi = \frac{H(-1/2\pi)}{\Psi_i(-1/2\pi)\Psi_s^*(-1/2\pi)} \tag{7}$$

$$= F_\xi\left(-\frac{4\pi}{\lambda}\right), \tag{8}$$

which encodes a particular spatial frequency of the backscattered illumination intensity. In this case $-1/2\pi$ is the frequency corresponding to the negative fundamental of the modulation waveform; the negative fundamental frequency is used because it encodes phase delay, rather than phase shift, so that the complex argument is proportional to range rather than negatively proportional. Equation 7 consists of a measurement of the negative fundamental Fourier bin of the measured correlation waveform, divided by the negative fundamental Fourier bin of the correlation waveform that would be expected in the ideal unity amplitude, zero range case. In the perfect single component return case the original amplitude and phase can be determined by

$$a_0 = |\xi| \tag{9}$$

and

$$r_0 = \frac{4\pi}{\lambda}\arg(\xi) + \frac{2}{\lambda}m, \tag{10}$$

where $m \in \mathbb{Z}$ is an arbitrary constant. To avoid explicit discussion of this cyclic ambiguity, this paper typically considers distance notated as phase rather than true range.

The most common approach to sampling the correlation waveform is the four differential phase step homodyne approach. In this case, an estimated complex domain range measurement is given by

$$\tilde{\xi} = \frac{\pi}{2\Psi_s(-1/2\pi)\Psi_i^*(-1/2\pi)}(h(0) - h(\pi) + j(h(\pi/2) - h(3\pi/2)), \tag{11}$$

where $h(0), h(\pi), h(\pi/2)$ and $h(3\pi/2)$ are each discrete differential measurements of the correlation waveform. This non-ideal sampling process is subject to systematic errors such as aliasing induced non-linearity [3] which are not considered further.

2.2 Modelling Mixed Pixels

While the single component return assumption is useful, in practice pixels at object boundaries integrate over more than one backscattering source. These are called mixed pixels.

The mixed pixel problem is not specific to range imaging. It may occur in any imaging science, where a pixel or integration/sampling region contains data from more than one discrete source. For example, Chang et al. [4] discusses the problem of decomposing mixed pixels in multispectral/hyperspectral images as linear combinations of discrete signatures. One possible application is the determination of precise land use statistics from LandSat images, despite significant quantities of mixed pixels. The difference between the LandSat decomposition problem and the AMCW decomposition problem is that each component return within a mixed AMCW measurement is composed of continuously variable amplitude and range values.

Mixed pixels are problematic for non-AMCW lidar systems as well: while range-gating [5] and full-waveform systems [6] are capable for the most part of separating out multiple returns within each pixel, if two returns are very close to each other then they cannot be separated. Erroneous range values due to multiple backscattering sources have been demonstrated in data captured using a pulsed system [7, 8].

One model for the backscattering function in the case of mixed pixel/multipath interference is a sparse spike train [9]: the sum of scaled and translated Dirac delta functions. In this case the backscattered signal intensity as a function of range is modelled as

$$f_{\xi}(r) = \sum_{i=0}^{M-1} a_i \delta(r - r_i) \tag{12}$$

$$F_{\xi}(u) = \sum_{i=0}^{M-1} a_i e^{-2\pi j u r_i}. \tag{13}$$

This allows the range measurement to be written as the sum of backscattering components, $\eta_i \in \mathbb{C}$, within the pixel measured at some modulation wavelength λ, viz

$$\xi - \sum_{i=0}^{M-1} \eta_i. \tag{14}$$

2.3 Perturbations Due to Multiple Returns

The simplest possible mixed case occurs when a primary return with an amplitude of one and a phase of zero is perturbed by a secondary return with a relative amplitude of b and a relative phase of θ. This case is modelled by the function

$$\Lambda_f(b, \theta) = 1 + b e^{j\theta}. \tag{15}$$

Assuming that the primary component/first term is the intended subject and that $\theta \neq 0$, as the relative amplitude of the secondary component increases $\Lambda_f(b, \theta)$ diverges from one. As a result, $\Lambda_f(b, \theta)$ can be considered to be the perturbation of the primary component return by the secondary component return. The behaviour of $\Lambda_f(b, \theta)$ is demonstrated in Fig. 6. Similar to aliasing perturbation, the amplitude and phase perturbations are ninety degrees out of phase.

One of the most frustrating aspects of multipath interference is that objects outside the field of view can scatter light onto the sensor; this can surreptitiously influence measurements such as phase/amplitude linearity calibrations.[1]

[1] Although primarily intended to mitigate aliasing of correlation waveform harmonics, these linearity calibrations also allow correction for systematic errors due to other effects like crosstalk.

Fig. 6 Phase and amplitude perturbation as a function of relative phase, θ, and relative amplitude, b. **a** Phase perturbation, $\arg(\Lambda_f(b,\theta))$. **b** Perturbed amplitude, $|\Lambda_f(b,\theta)|$

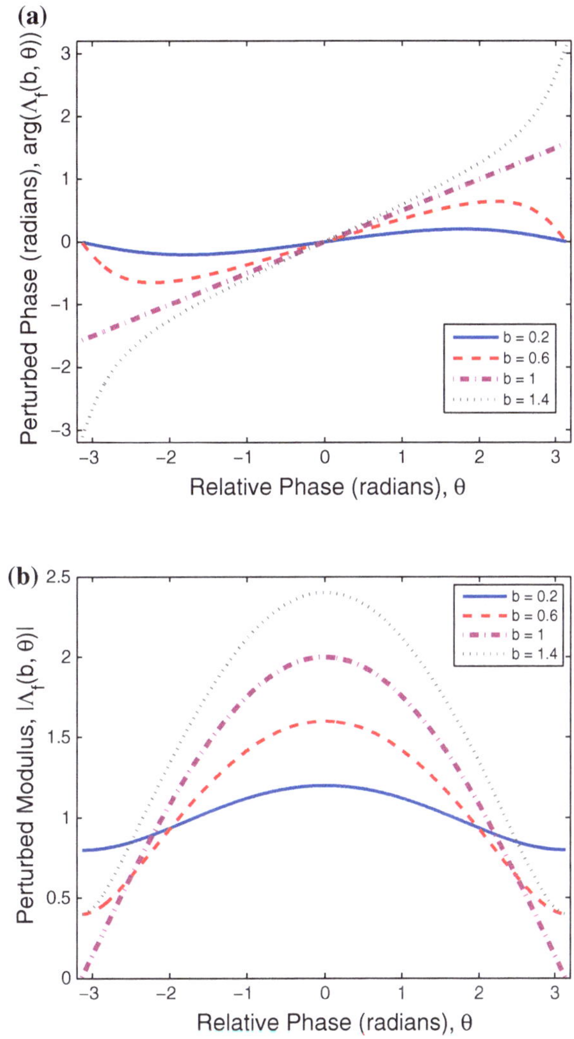

A standard approach to calibration for range-dependent phase and amplitude errors is to take measurements at different distances from the camera using a translation stage. As a practical example of multipath, we now model the impact of scattered light on a linearity calibration using a translation stage in the same manner as [3]. Because the relative amplitude of the scattering light increases as the translation stage moves away from the camera, the resultant linearity curve shape can be quite complicated. One possible model is

$$\xi = \frac{a_0}{d_0^2} e^{4\pi j d_0/\lambda} + a_1 e^{4\pi j d_1/\lambda}, \tag{16}$$

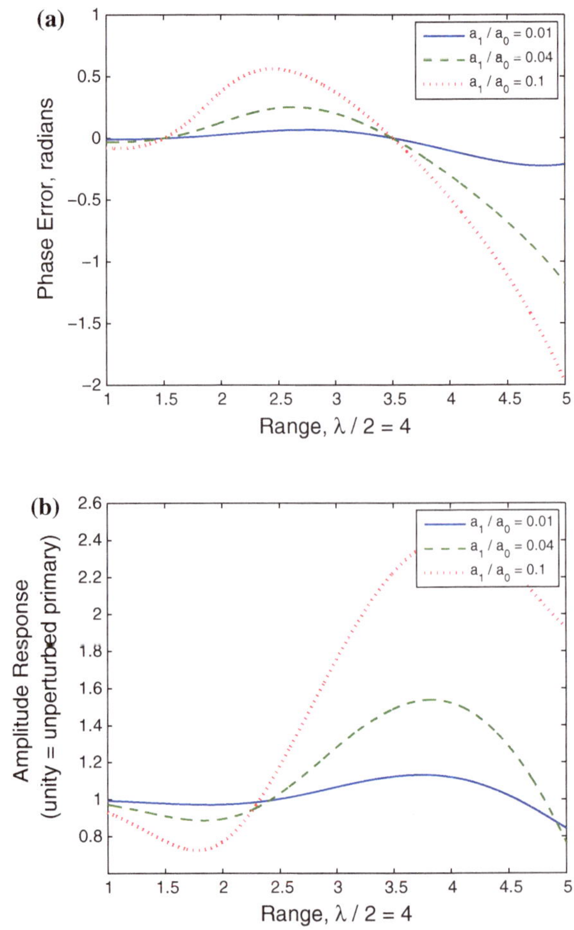

Fig. 7 A model of the phase and amplitude perturbations introduced by mixed pixels for a translation stage, as a function of the relative amplitude of the perturbing scattered light at a 1 m distance using Eq. 16. **a** Phase perturbation. **b** Perturbed amplitude

where a_1 and d_1 are the amplitude and range to a static, perturbing return and a_0 is the amplitude of the translation stage return at a range $d_0 = 1$, where the amplitude is assumed to decay with the inverse squared law. An example is plotted in Fig. 7. As a general rule, if a linearity calibration is not cyclic over each ambiguity interval, it is probably due to either temperature drift or multipath interference.

3 Point Scanning Systems and Detection Approaches

There are a number of existing detection and amelioration approaches for mixed pixels/multipath interference reported in the literature. The first instance of AMCW mixed pixels in the literature appears to occur in the early nineties with regards to point scanning systems [10–12]. Much of the more recent research into full-field multipath appears unaware of this early work, perhaps partly due to the choice of terminology and change in technology (for example, [13, 14]). We begin by addressing this early work and then progress to full-field specific research.

Kweon et al. [10] deal with the experimental characterisation of a point scanning AMCW system and makes the first known (passing) reference to the AMCW lidar mixed pixel problem. Nevertheless, probably the most widely referenced early paper discussing mixed pixels is that of Hebert and Krotkov [11]. They present a detailed characterisation of the systematic errors that point scanning systems are subject to, such as distortion due to scanning and temperature induced temporal measurement drift. As part of this analysis, the mixed pixel and range-intensity crosstalk problems are introduced. While Hebert and Krotkov suggest that the electrical design and dynamic range/detector saturation issues are responsible for range-intensity crosstalk, depending on the precise detector design it is possible that multipath interference could be a contributing factor. Hebert and Krotov explain the fundamental formation process for mixed pixels including the symptoms such as non-interpolative/wraparound range estimates due to $|\theta| > \pi$.

As a solution to the mixed pixel problem Hebert and Krotov [11] suggest the application of median filters; unfortunately in practice, this is a limited fix. While wraparound blurring results in extreme range measurements, also referred to by some authors as 'flying pixels' [15], that are removed by a median filtering process, interpolative blurring when $|\theta| < \pi$ results in intermediate range estimates which are not necessarily removed. Later work has described median filtering as 'failing catastrophically' in many cases [16]. Reliable mixed pixel detection/removal requires more advanced techniques.

3.1 Normal Angle and Edge Length Detection

There are a number of quite simple approaches suitable for mixed pixel detection; Tang et al. [17] provides a performance comparison of several related methods. They utilise a 2D matrix of 3D coordinates, thus retaining elementary connectivity information. Generating a 3D mesh and removing edges that are detected as discontinuities, any isolated pixels can be considered to be erroneous values. Tang et al. applies three discontinuity detection methods: the first thresholds edge length. In general, triangles formed between mixed pixels and neighbouring unperturbed pixels have longer edge lengths. Thus thresholding edge length allows detection of improbable surfaces. The second method thresholds the normal angle of triangles; in general, even if pixels are not strictly mixed, the range information measured for surfaces with a high angle of incidence is poor, so removal of these points is often advantageous. One obvious limitation is that subtly mixed pixels cannot be detected with these approaches.

The third algorithm is based around detecting specific pixels, rather than triangles at discontinuities. The number of adjacent pixels that fall within a specified angular domain relative to the pixel in question are counted. An explanatory diagram is given in Tang et al. [17]. While the algorithm is restricted to the eight-connected neighbourhood of the pixel using the known connectivity information, a similar approach could be extended to a wider neighbourhood. In general, Tang et al. found that none of the methods were completely reliable, but that the normal angle method performed best. In practice, normal angle detection appears to be one of the most common approaches. The downside is that cleaning data of all mixed pixels can result in systematic errors, such as underestimation of column widths in architectural modelling [18]. As with all these detection methods, there is a difficult trade-off between sensitivity and specificity.

3.2 Adam's Detection Algorithm

Adams/Adams and Probert [12, 16] developed a method for detecting range and reflectance discontinuities in point scanning systems, although only the range discontinuities typically result in erroneous range estimates. Simplifying his notation, for a range measurement given by a linear interpolation between two backscattering sources, namely

$$\xi = (1 - \rho)\eta_0 + \rho\eta_1, \tag{17}$$

where $\rho \in [0, 1]$ is an interpolation constant, the value s given by the derivative

$$s = \frac{d^2|\xi|^2}{d\rho^2}, \tag{18}$$

is a fixed constant, calculable by

$$s = 2|\eta_0|(1 + b^2 - 2b\cos(\theta)), \tag{19}$$

where $b = |\eta_1|/|\eta_0|$. By modelling the manner in which the region integrated by the circular cross-section of the illumination changes as the point is scanned across the scene, Adams numerically determined and thresholded s, thus identifying both range and reflectance discontinuities. Unfortunately, this is a pretty difficult process to implement and most scanners are designed to produce discrete independent measurements, rather than continuously varying overlapped measurements. In theory a very similar algorithm could potentially be applied to images subject to defocus limitations by modelling the defocus point spread function (PSF) as integrating over a region in a manner similar to the point scanner illumination cross-section.

Another detection approach by Adams [19] involves using a Kalman filter to predict range values for a scanning system, which allows for the detection of statistically significant edges. By applying this to range data, potentially mixed pixels near edges can be discarded. Other work on general edge detection in range-data includes that of Sappa and Devy [20] in the context of image segmentation.

Because of the substantial body of literature on edge detection and segmentation, we do not address these approaches any further.

3.3 Classification Based Detection

An approach demonstrated by Tuley et al. [7, 8] involves classifying points in a point cloud within particular regions of interest. By performing Principal Components Analysis on the points within the region of interest, it is possible to classify the points as being either a surface, a linear formation or randomly distributed; a classification approach first demonstrated by Vandapel et al. [21]. For eigenvalues of the covariance matrix, $\lambda_0 \geq \lambda_1 \geq \lambda_2$, it was found that $\lambda_0, \lambda_1 \gg \lambda_2$ generally indicates a surface, $\lambda_0 \gg \lambda_1, \lambda_2$ indicates a linear formation and $\lambda_0 \approx \lambda_1 \approx \lambda_2$ indicates a disorganised group of points. By training a classifier on hand-labelled data, surfaceness, linearness and randomness metrics were automatically calculated [21]. Choosing regions with relatively low angular width, Tuley et al. [7, 8] found high surfaceness and surface normals near orthogonal to the camera to be useful indicators of mixed pixels. The primary advantage of this method appears to be the applicability to pure point clouds, rather than solely 2D matrices of 3D Cartesian points.

3.4 Characterising Error at Spatial Discontinuities

Tang et al. [18] characterise the systematic error in measurements of object size/width using point scanning systems, although the concepts can be easily extended to the full-field case. Tang et al. [18] analyses two different types of point scanner: first response pulsed[2] and AMCW systems. Given knowledge of the spot size and the spacing between spots, it is possible to place bounds on the accuracy of the estimated width or height of a scanned object, such as a column: this is extremely important for engineering metrology applications. Let the angular width of the spot for sample i be notated as $\theta_w(i)$ and the location of each spot be $\theta_s(i)$. As scanning in a line across an object such as a pillar, if a non-mixed foreground measurement, i, is followed immediately by a non-mixed background measurement, $i+1$, then the location of the transition between the two is known to exist at a location β, bounded by

$$\theta_s(i) + \frac{\theta_w(i)}{2} < \beta < \theta_s(i+1) - \frac{\theta_w(i+1)}{2}. \tag{20}$$

[2] In other words, a pulsed system which only returns the range to the closest backscattered return.

However, if the first measurement is known to be mixed then the transition must occur within the region measured by sample i, giving

$$\theta_s(i) - \frac{\theta_w(i)}{2} < \beta < \theta_s(i) + \frac{\theta_w(i)}{2}. \tag{21}$$

Depending on how the scene is sampled and the sensitivity of the mixed pixel detection method there are a number of possible enhancements to the model; for example, Tang et al. models the impact of angle of incidence. First response pulsed systems have very different behaviour: assuming that the foreground return is bright enough to be detected, then without mixed pixel detection the true boundary could occur either within sample i or in the void between samples, that is

$$\theta_s(i) - \frac{\theta_w(i)}{2} < \beta < \theta_s(i+1) - \frac{\theta_w(i+1)}{2}. \tag{22}$$

4 Restoration and Full-Field Systems

Because of the subtle perturbations that mixed pixels introduce into range-images, the existence of an intensity-coupled range error in full-field systems has been known for a long time, although it has generally not been connected to the mixed pixel problem in point-scanning systems. A number of authors have attempted to calibrate for these errors without fully characterising the origin. For example, Oprisescu et al. [13] identified an intensity related distance error and attempted a fixed calibration using the assumption that the calibration is scene independent. Because of the lack of further characterisation, it is difficult to know whether the authors have ended up with a global characterisation for mixed pixel effects, despite the scene dependence, or whether they could have possibly conflated distance as a function of amplitude with distance and amplitude and a function of distance, due to the impacts of aliasing on both amplitude and phase. It is notable that the paper does not mention aliasing at all.

A more advanced calibration was performed by Lindner et al. [15, 22]. He calibrated for radial distortion, perspective projection, aliasing non-linearity and reflectivity induced errors, the latter using B-Splines. Like Oprisescu, Lindner does not posit a cause for the reflectivity based errors. A similar calibration was performed by Abdo et al. [23] using a lookup table.

Gudmundsson et al. [14] was one of the first papers to identify the impact of scene structure on range measurements. Using an example of the corner of a room, he found that the measured surfaces became non-planar as the light scattered off the neighbouring surfaces resulted in systematic overestimation of range: a clear case of intra-scene multipath interference, although no attempt was made to restore the systematic error. Other work, such as that of Karel et al. [24] and Falie et al. [25] have sought to characterise the impact of intra-camera scattering, due to subtle scattering and reflection effects. Karel et al. [24] experimentally demonstrated that the amplitude of the scattering point spread function (PSF) was independent of range[3] and that the PSF is generally rotationally symmetric around the principal point.

Another paper, by Jamtsho and Lichti [26] empirically characterises intra-lens scattering and then applies two different, albeit fairly limited, spline-based fixed-calibration compensation methods using SwissRanger imagers (SR-3000 and SR-4000).

4.1 Scattering Restoration Using Scene Texture, Structured Light or Calibration Squares

One approach to removing the perturbations from scattered light from range images is to use texture or patterning in the scene. Godbaz et al. [27] present an algorithm using scene texture to make local estimates of scattered light.

In general, scattering PSFs in practical range-imaging systems are large and the scattered light can be modelled as locally homogeneous. Using the approach of Godbaz et al. [27] for a fixed perturbation, $\lambda \in \mathbb{C}$, the complex domain measurement at any pixel can be written as

$$\xi = \lambda + \eta \tag{23}$$

$$= \lambda + ae^{j\theta}. \tag{24}$$

If range—hence phase, θ—is assumed to be fixed over the local region, for example a surface orthogonal to the camera, then Eq. 24 can be rewritten as

$$\Re(\xi) = \Re(\lambda) + a\cos(\theta) \tag{25}$$

$$\Im(\xi) = \Im(\lambda) + a\sin(\theta), \tag{26}$$

where $\Re(x)$ and $\Im(x)$ are the real and imaginary operators. Equations 25 and 26 can be rearranged to give

$$\Im(\xi) = \tan(\theta + \pi N)(\Re(\xi) - \Re(\lambda)) + \Im(\lambda) \tag{27}$$

$$= \alpha + \beta\Re(\xi), \tag{28}$$

where $\alpha \in \mathbb{R}$ encodes the influence of the scattered light upon the data, $\beta \in \mathbb{R}$ encodes the phase of the primary return and $N \in \mathbb{Z}$ is a disambiguation constant

[3] Strictly, the scattering PSF is complex domain, and the phase of the complex number is a linear function of range. It is important to distinguish the scattering PSF, which is independent of range, from the defocus PSF, which is not.

for the tangent function. Determination of α and β corresponds to trying to fit a linear model to the data, where the imaginary parts of the range measurements are modelled as a function of the real parts. In the ideal unperturbed case, $\alpha = 0$, and the line goes through the origin. As the amount of scattered light increases, the line is usually perturbed in such a manner that it no-longer passes through the origin, resulting in erroneous estimates of phase.

Godbaz et al. [27] fit Eq. 28 simultaneously across the entire image for each pixel using a Fourier implementation of linear least squares. Given values of β for each pixel it is possible to estimate the phase for each pixel using texture. If the mean signal intensity (including signal and the offset from ambient light) is known, then it is also possible to remove the phase ambiguity and determine N. If χ is the mean signal intensity, then disambiguated phase estimates can be made by calculating the covariance of χ with the real part of ξ, viz

$$\kappa = cov(\chi, \Re(\xi)) \tag{29}$$

$$\hat{\theta} = atan2(\beta\kappa, \kappa), \tag{30}$$

giving an estimate of the true underlying phase, $\hat{\theta}$.

It is also possible to estimate the scattered light, λ, at each pixel. Whereas Eq. 30 operates by using a region containing (hopefully) only a single object, by combining multiple estimates of α and β from objects at different ranges, determination of λ can be achieved by an additional linear fit. The relationship between values of α, β and the estimate $\hat{\lambda}$ is found by fitting

$$\Im(\hat{\lambda}) + \Re(\hat{\lambda})\beta = \alpha, \tag{31}$$

over a large spatial region.

It is not necessary for scenes to be highly textured: for example, Falie and Buzuloiu [28, 29] used calibration squares with known reflectivities on objects in the scene to remove the impact of any relatively homogeneous scattering. In the simplest case, given a known, perfectly non-reflective calibration square, a measurement of the square could simply be subtracted from the entire image.

Another related approach is to use structured illumination, as first suggested by Falie [30]. This approach is aimed primarily at intra-camera scattering, although it could be theoretically applied to certain types of intra-scene multipath interference in limited situations; it is definitely not suitable for mixed pixels. Using a second modulated illumination source to provide additional illumination of a small subregion of the camera field-of-view, Falie [31] found that it is possible to determine both the correct range measurements and the perturbing measurement for the subregion of the image. The approach relied upon the magnitude of the correct return increasing in a known manner with the addition of a second illumination source, while the light scattered from other parts of the scene remained relatively constant. Whereas Godbaz et al. [27] relied upon spatial variation in intensity, Falie [31] achieved the same goals using temporal variation in intensity over a subregion of the image; these are closely related approaches. While the former is fundamentally limited by the high frequency spatial content of the range-image, the latter could be extended to dense sampling of the scattered light using pattern projection equipment, for example a micromirror array or LCD screen, although this has yet to be demonstrated.

4.2 Restoration by Spatial Deconvolution

Rather than attempting to deduce the nature of the scattering within the scene by analysing the variation in complex domain range measurements, it is also possible to directly model the scattering process as the convolution of an image with a PSF. By structuring the intra-camera scattering problem as a deconvolution problem, it is possible to apply off-the-shelf deconvolution algorithms in order to reconstruct a less perturbed scene. In actuality, the restorations can only be partial at best, due to factors such as light scattered from outside the field-of-view and difficulties involved in modelling the full spatial variance of the scattering PSF.

One of the biggest difficulties is simply measuring the PSF in an off-the-shelf ranging system. While some authors [32] have used have used retroreflective dots, others have been forced to resort to blind estimation of the PSF by an expert [33, 34]. Ideally, a perfect point source should be used for PSF estimation. In practice, high quality measurements require a second illumination source other than the array of LEDs utilised by the vast majority of commercial systems. Because most systems use differential measurements, which are designed to cancel out ambient light, it is not possible to use a non-modulated point source. Godbaz et al. [27] used a laser ducted down an optical fibre as a point source for a custom-built system.

In general, one can model the measured range values as a linear transformation of the underlying true complex phasors, viz

$$\begin{pmatrix} \xi_0 \\ \xi_1 \\ \vdots \\ \xi_{n-1} \end{pmatrix} = H \begin{pmatrix} \eta_0 \\ \eta_1 \\ \vdots \\ \eta_{n-1} \end{pmatrix}. \tag{32}$$

In the isoplanatic (non-spatially variant PSF) case, H is a block Toeplitz matrix and the vector of unperturbed measurements can generally be solved for in the Fourier domain, depending on the assumed prior distribution.[4] However, in practice, the PSF changes greatly across the field-of-view. Anisoplanatic (spatially varying) PSFs generally require the application of iterative techniques, such as the

[4] For example, from a Bayesian perspective Tikhonov regularisation corresponds to the assumption of a Gaussian distribution of intensity values.

Landweber [35] and Lucy-Richardson [36, 37] algorithms. Typically, this results in a significant increase in computational complexity. It has not yet been proven whether there is a significant improvement provided by modelling the spatial-variance, although this is likely to be scene dependent.

Mure-Dubois and Hügli [33, 34] implemented a restoration method capable of operating in real-time. This was achieved by designing an special inverse filter. They modelled the PSF as the sum of several 2D Gaussian distributions, each with different width parameters and weightings. Through clever choice of parameters, it is possible to model such a PSF as separable convolutions: one vertical and one horizontal. The authors demonstrated the algorithm on data produced by a Mesa SR-3000, where light from a foreground object was scattered onto a dark background, resulting in erroneous range measurements.

Kavli et al. [32], developed a restoration method for removing the impact of scattering on an image using the assumption of an anisoplanatic scattering function and a simple iterative method. The proposed method was demonstrated on a SwissRanger SR-3000, sampling the scattering PSF using retroreflective dots. The algorithm was clearly shown to mitigate range-intensity coupling in example scenes.

Because mixed pixels can be caused by limited depth-of-field, one mitigation approach is to use the range data to determine defocus blur scale and perform a spatially variant deconvolution to refocus the image and sharpen the object transitions. Godbaz et al. [2] used a coded aperture to increase the bandwidth of the modulation transfer function more broadband and demonstrate a proof-of-concept restoration algorithm.

4.3 Restoration by Modelling Intra-Scene Scattering

Fuch [38] developed what is probably the most advanced multipath interference model in the current literature. Intra-scene multipath reflections are modelled within the FOV by assuming that each surface in the FOV is a Lambertian reflector and scatters light onto all the other objects in the scene. Using the perturbed range and amplitude measurements it is possible to estimate the multipath perturbation for each pixel and subtract it from the range-image.

There are several significant flaws with this approach, including the use of perturbed range data, which may result in erroneous estimates of the sum reflected light. Even using an iterative approach, where the corrected range-data is used to reestimate the scattered light, convergence is not necessarily guaranteed, especially for highly perturbed points. Other issues include specular reflection, reflections and scattered light from objects outside the camera FOV and execution time (approximately ten minutes per image for the reported implementation). There is no discussion of occlusion in the paper, so accurate restorations are most probably limited to simple scene configurations like the internal corners of rooms; a more advanced approach would probably require ray tracing. Despite these issues it is an interesting and original approach, but it does not have the utility of, for example, the intra-camera scattering deconvolution approaches.

4.4 Mixed Pixel Restoration by Parametric Surface Modelling

One of the biggest problems with the mixed pixel algorithms most commonly applied to point scanners is that they generally detect mixed pixels and throw them away. Because full-field systems integrate over large spatial regions there is an increased occurrence of mixed pixels, particular when one factors in limited depth-of-field. Larkins et al. [39] developed a new approach that detects mixed pixels using a variation of the normal-angle filter described in Sect. 3.1. Once all the mixed pixels in the image have been detected, Larkins et al. thresholds the neighbouring pixels into two groups, being a foreground and background object, using the Otsu method [40]. A bivariate second order polynomial is then fit to each of the classified points using a linear least squares approach in order to model the shape of the surface. Each pixel is then projected onto the closest surface to estimate corrected values.

This sort of projection approach can be considered to be an improvement over the simple mixed pixel removal approaches, however like all restoration algorithms runs the risk of introducing other errors depending on the particular shape of objects in the scene.

4.5 Correlation Waveform Deconvolution/Waveform Shape Fitting

Given that each AMCW range measurement is equivalent to sampling a particular spatial frequency of the backscattered signal returns, by taking large sequences of measurements at different modulation frequencies it is possible to recover a model for the signal returns. Simpson et al. [41] achieved this by taking a sequence of 20 measurements between 10 and 200 MHz in an harmonic sequence; using an inverse Fourier transform it was possible to recover an extremely low resolution model. While this frequency stepped AMCW technique is highly limited due to the extremely large number of measurements and the modulation frequency bandwidth required, this approach nevertheless suggests that taking multiple measurements at different modulation frequencies may be the key to separating out different backscattering sources. In this section we discuss the implicit sampling of multiple modulation frequencies using the harmonics of the correlation waveform; in the next section we discuss the explicit sampling of different modulation frequencies.

Whereas a single isolated complex domain range measurement is only able to represent the properties of a single backscattering source, or at best the relationship between two backscattering components, by analysing the correlation waveform in greater depth it is possible to separate out multiple backscattering components within each pixel.

From Eq. 6 the measured correlation waveform is a convolution with a reference waveform, $\psi = \psi_i \star \psi_s$, where $\star$ represents correlation. In order for a frequency to be present in the reference waveform, it must be present in both $\Psi_i(u)$ and $\Psi_s(u)$. From a spatial frequency content perspective, if the reference waveform contains harmonics other than the fundamental then every AMCW lidar measurement is implicitly sampling multiple spatial frequencies of the backscattered signal returns.[5] Thus one can infer a lot more about the backscattering sources from the correlation waveform than is typically assumed. This means that like FMCW and range-gating/pulsed measurements, in certain circumstances it is possible to separate out multiple returns. In practice, while the reference waveform is relatively band-limited, there is enough harmonic content to pose the multiple return separation problem as a deconvolution problem. That is, determination of $f_\xi(r)$, when given $f_\xi * \psi_i \star \psi_s$.

Perhaps the greatest parallels are in full-waveform lidar. A significant amount of research has been performed into fitting models in full-waveform lidar: for example, Hofton et al. [42] and Chauve et al. [43] use numerical methods to fit Gaussian and generalised Gaussian models respectively to the measured waveform. Stilla and Jutzi [6] discuss a variety of different techniques, including: additional models for the waveform pulse, such as rectangular functions; analysis of component returns, parameterising them by time, width and amplitude; component return detection methods, including peak detection, leading edge detection, constant fraction detection, centre of gravity detection; and the application of deconvolution methods. The main problem with the most common approaches to fitting distributions to recorded waveforms is that they necessitate iterative solutions, which are highly computationally intensive when performed across a 2D matrix of measurements. Also, the majority of the models rely on the shape being a nicely-defined continuous function, like a Gaussian distribution; the correlation waveform is often better modelled by a piecewise model, such as a truncated triangle waveform [44].

Another technique, often utilised for the processing of FMCW range-data, is matched filter processing [45, 46]. However, matched filter processing does not reconstruct missing spatial frequencies in the same manner that well-designed

[5] This manifests as aliasing if only a small number of samples are taken of the correlation waveform, hence the importance of harmonic cancellation techniques.

deconvolution methods do. From the point-of-view of a typical AMCW correlation waveform, application of a matched filter to estimate $f_\xi(r)$ is utterly useless.

A number of attempts have been made to use the harmonic content of the correlation waveform to separate out multiple returns. Godbaz et al. [9] poses the problem as a sparse spike train deconvolution problem, implicitly posing the problem as a discrete variation of Eqs. 12 and 13. By applying the Levy-Fullagar deconvolution algorithm [47] it was possible to separate out multiple components within each pixel; the method was also found to improve precision by 30 % in the single return case. However, from a practical perspective this is not an advantageous approach because an AMCW waveform is not designed for harmonic content; in fact most systems try deliberately to minimise it. As a result, long integration times and high data rates were required. If one is willing to accept those, then specialist modulation techniques like FMCW and range-gating are probably more worthwhile.

Other work attempts to directly fit the correlation waveform shape. Godbaz et al. [44] fit two different piecewise models to the correlation waveform shape: a truncated triangle model derived from the correlation of two rectangular function and another model based on linear interpolation between two known basis vectors. The models were fit using a Poisson maximum likelihood numerical optimisation method and the separation of components within a pixel was demonstrated.

4.6 Multifrequency Methods

Whereas the correlation waveform models in the second above implicitly sampled multiple spatial frequencies of the signal returns, it is also possible to explicitly sample different spatial frequencies of the signal returns by taking several sequential measurements at different modulation frequencies. In the simplest case this involves taking two measurements with a frequency ratio of 2:1 in order to separate out two component returns.

For two coprime relative frequencies r_0 and r_1, where the underlying component returns, $\eta_0, \eta_1 \in \mathbb{C}$ are notated at a relative frequency of one,[6] the measurements ξ_0 and ξ_1 can be written in terms of η_0 and η_1 as

$$\xi_0 = \frac{\eta_0^{r_0}}{|\eta_0|^{r_0-1}} + \frac{\eta_1^{r_0}}{|\eta_1|^{r_0-1}} \tag{33}$$

$$\xi_1 = \frac{\eta_0^{r_1}}{|\eta_0|^{r_1-1}} + \frac{\eta_1^{r_1}}{|\eta_1|^{r_1-1}}. \tag{34}$$

[6] For example, if using frequencies of 30 and 20 MHz, the component returns, η_0 and η_1, would be notated as if captured at a frequency of 10 MHz. This means that there is no cyclic ambiguity, and enables the representation in Eqs. 33 and 34.

This corresponds to an extremely non-linear optimisation problem and has no obvious trivial closed-form solution. However, using numerical optimisation techniques it is possible to separate out the different components within each pixel. Dorrington et al. [48] demonstrate this using the Mesa Imaging SR-4000 and Canesta XZ-422 cameras for a frequency ratio of 2:1, with encouraging results. Nevertheless, this technique remains highly experimental.

5 Conclusion

While mixed pixels are relatively well-understood and can be easily removed from range-images, the multipath interference problem remains fundamentally unsolved. Currently, intra-camera scattering can be adequately compensated for by use of deconvolution techniques, however intra-scene scattering is far more challenging. In terms of future potential, attempting to model scene structure, while interesting, appears to be a fundamentally limited approach. It is too easy for light to be scattered from outside the field-of-view of the camera. Perhaps one of the more useful potential innovations would be a reliable method to determine the degree of perturbation, thus place accurate bounds on the true value of any measurements.

Functional and practical intra-scene multipath compensation may require the use of more advanced modulation techniques, possibly simultaneous capture of multiple frequencies. Ultimately as technology improves, FMCW and range-gating techniques may become better options, as has occurred for radar. For the moment, despite claims of high precision, full-field AMCW lidar systems remain fundamentally accuracy limited and it is quite important for the experimenter to be aware of the fundamental limitations of the technology in order to achieve reliable results.

6 Understanding Mixed Pixels and Multipath Interference

6.1 Lidar Modulation Techniques

Multipath interference is a more general term than the mixed pixel problem and refers to the broader case including intra-scene reflections. For example, when imaging the corner of a room it is common for light to be reflected several times, resulting in blurry range-data. Multipath interference is also caused by reflections within the optics of full-field range-imagers, resulting in crosstalk between measurements.

References

1. A.A. Dorrington, M.J. Cree, A.D. Payne, R.M. Conroy, D.A. Carnegie, Meas. Sci. Tech. **18**(9), 2809 (2007)
2. J.P. Godbaz, M.J. Cree, A.A. Dorrington, in *ACCV*, 2011. Lecture Notes in Computer Science, vol. 6495 (Springer, Berlin, 2011), pp. 397–409
3. J.P. Godbaz, M.J. Cree, A.A. Dorrington, Remote Sens. **4**, 21–42 (2012)
4. C.I. Chang, X.L. Zhao, M.L.G. Althouse, J.J. Pan, IEEE Trans. Geosci. Remote Sens. **36**, 898 (1998). doi:10.1109/36.673681
5. B.W. Schilling, D.N. Barr, G.C. Templeton, L.J. Mizerka, C.W. Trussell, Appl. Opt. **41**, 2791 (2002)
6. U. Stilla, B. Jutzi, in *Topographic Laser Ranging and Scanning. Principles and Processing*, ed. by J. Shan, Chap. 7 (CRC Press, Boca Raton, 2009), pp. 215–234
7. J. Tuley, N. Vandapel, M. Hebert, Analysis and removal of artifacts in 3-d ladar data. Technical report, Carnegie Mellon University Robotics Institute, 2004
8. J. Tuley, N. Vandapel, M. Hebert, in *IEEE International Conference on Robotics and Automation*, 2005, pp. 2203–2210
9. J.P. Godbaz, M.J. Cree, A.A. Dorrington, in *Image and Vision Computing New Zealand (IVCNZ'08)*, 2008, pp. 1–6
10. I.S. Kweon, R. Hoffman, E. Krotkov, Experimental characterization of the perceptron laser rangefinder. Technical report, Robotics Institute Carnegie Mellon University, 1991
11. M. Hebert, E. Krotkov, IVC **10**, 170 (1992)
12. M.D. Adams, in *Proceedings of IEEE International Conference on Robotics and Automation*, vol. 2, 1993, pp. 8–13
13. S. Oprisescu, D. Falie, M. Ciuc, V. Buzuloiu, in *Proceedings of International Symposium on Signals, Circuits and Systems 2007, ISSCS*, 2007
14. S.A. Guomundsson, H. Aanaes, R. Larsen, in *Proceedings of International Symposium on Signals, Circuits and Systems 2007, ISSCS*, 2007
15. M. Lindner, I. Schiller, A. Kolb, R. Koch, Comput. Vis. Image Underst. **114**, 1318 (2010). doi:10.1016/j.cviu.2009.11.002
16. M.D. Adams, P.J. Probert, Int. J. Rob. Res. **15**(5), 441 (1996)
17. P. Tang, D. Huber, B. Akinci, *A Comparative Analysis of Depth-Discontinuity and Mixed-Pixel Detection Algorithms* in *Proceedings of the International Conference on 3-D Digital Imaging and Modeling (3DIM)* (Los Alamitos, 2007), pp. 29–38
18. P. Tang, B. Akinci, D. Huber, Quantification of edge loss of laser scanned data at spatial discontinuities. Autom. Constr. **18**, 1070–1083 (2009)
19. M.D. Adams, in *Proceedings of 2001 IEEE/RSJ International Conference on Robots and Systems*, 2001, pp. 1726–1731
20. A.D. Sappa, M. Devy, in *Proceedings of Third International Conference on 3-D Imaging and Modelling (3DIM '01)*, 2001
21. N. Vandapel, D.F. Huber, A. Kapuria, M. Hebert, in *Proceedings of the IEEE International Conference on Robotics and Automation ICRA '04*, vol. 5, 2004, pp. 5117–5122
22. M. Lindner, A. Kolb, in *Proceedings of the SPIE Intelligent Robots and Computer Vision XXV: Algorithms, Techniques, and Active Vision*, vol. 6764, 2007. doi:10.1117/12.752808
23. N. Abdo, A. Borgeat, 3D camera calibration. Technical report, 2010
24. W. Karel, S. Ghuffar, N. Pfeifer, in *International Archives of Photogrammetry, Remote Sensing and Spatial Information Sciences, Vol. XXXVIII, Part 5 Commision V Symposium*, 2010
25. D. Falie, V. Buzuloiu, in *Proceedings of 2007 International Symposium on Signals, Circuits and Systems (ISSCS)*, 2007
26. S. Jamtsho, D. Lichti, in *Proceedings of the ISPRS Commission V Mid-Term Symposium on Close Range Image, Measurement Techniques*, 2010

27. J.P. Godbaz, M.J. Cree, A.A. Dorrington, in *Image and Vision Computing New Zealand 2009 (IVCNZ' 09)*, 2009, pp. 304–309
28. D. Falie, V. Buzuloiu, in *Proceedings of the IEEE International Workshop on Imaging Systems and Techniques, IST* 2008
29. D. Falie, Improvements of the 3D images captured with Time-Of-Flight cameras. Technical report, Politehnica University of Bucharest, 2009
30. D. Falie, in *2008 International Conference on Optical Instruments and Technology: Optical Systems and Optoelectronic Instruments. Proceedings of the SPIE*, vol. 7156, 2008. doi:10.1117/12.807125
31. D. Falie, IET Image Process. **5**(5), 523 (2011)
32. T. Kavli, T. Kirkhus, J.T. Thielemann, B. Jagielski, in *Two- and Three-Dimensional Methods for Inspection and Metrology VI (SPIE)*, 2008. doi:10.1117/12.791019
33. J. Mure-Dubois, H. Hügli, in *Two- and Three-Dimensional Methods for Inspection and Metrology V*, vol. 6762 (SPIE, Boston, 2007)
34. J. Mure-Dubois, H. Hügli, in *Proceedings of the ICVS Workshop on Camera Calibration Methods for Computer Vision Systems, CCMVS2007*, 2007
35. L. Landweber, Am. J. Math. **73**(3), 615 (1951)
36. L.B. Lucy, Astron. J. **79**, 745 (1974)
37. W.H. Richardson, J. Opt. Soc. Am. (1917–1983) **62**, 55 (1972)
38. S. Fuchs, in *Proceedings of the 2010 20th International Conference on Pattern Recognition*, (IEEE Computer Society, Washington, DC, 2010), pp. 3583–3586. doi:10.1109/ICPR.2010.874
39. R.L. Larkins, M.J. Cree, A.A. Dorrington, J.P. Godbaz, in *Image and Vision Computing New Zealand 2009 (IVCNZ '09)*, 2009, pp. 431–436
40. N. Otsu, IEEE transactions on systems. Man Cybern. **9**(1), 62 (1979). doi:10.1109/TSMC.1979.4310076
41. M.L. Simpson, M.D. Cheng, T.Q. Dam, K.E. Lenox, J.R. Price, J.M. Storey, E.A. Wachter, W.G. Fisher, Appl. Opt. **44**, 7210 (2005)
42. M.A. Hofton, J.B. Minster, J.B. Blair, IEEE Trans. Geosci. Remote Sens. **38**, 1989 (2000). doi:10.1109/36.851780
43. A. Chauve, C. Mallet, F. Bretar, S. Durrieu, M.P. Deseilligny, W. Puech, in *International Archives of Photogrammetry, Remote Sensing and Spatial Information Sciences*, 2007
44. J.P. Godbaz, M.J. Cree, A.A. Dorrington, in *SPIE 7251—Image Processing: Machine Vision Applications II*, vol. 7251 (SPIE, San Jose, 2009), p. 72510T
45. M. Soumekh, *Synthetic Aperture Radar Signal Processing with MATLAB Algorithms* (Wiley, New York, 1999)
46. M. Jankiraman, *Design of Multifrequency CW Radars* (Scitech, Raleigh, 2007)
47. S. Levy, P.K. Fullagar, Geophysics **46**(9), 1235 (1981)
48. A.A. Dorrington, J.P. Godbaz, M.J. Cree, A.D. Payne, L.V. Streeter, in *SPIE 7864—Three-Dimensional Imaging, Interaction, and Measurement*, San Francisco, 2011

3D Cameras: Errors, Calibration and Orientation

Nobert Pfeifer, Derek Lichti, Jan Böhm and Wilfried Karel

1 Introduction

In this chapter range cameras and especially the data they provide are investigated. Range cameras can be considered to provide a 3D point cloud, i.e. a set of points in 3D, for each frame. The overall aim of this chapter is to describe techniques that will provide point clouds that:

- Are "free" of systematic errors (definition to follow); and
- Are registered together from multiple images or an image stream into one superior coordinate system.

In order to achieve this goal, the quality of range camera data has to be analyzed. The error sources need to be studied and grouped into random and systematic errors. Systematic errors are errors that can be reproduced under the same measurement conditions, e.g. a range measurement may be systematically affected by the brightness of the scene. Random errors are independent of each other and of the scene. Their influence can be reduced by averaging. Systematic errors can be reduced if additional aspects of the measurement instrument or the scene are

N. Pfeifer · W. Karel (✉)
Department of Geodesy and Geoinformation, Vienna University of Technology, Vienna, Austria
e-mail: wilfried.karel@geo.tuwien.ac.at

N. Pfeifer
e-mail: norbert.pfeifer@geo.tuwien.ac.at

D. Lichti
Department of Geomatics Engineering, University of Calgary, Calgary, Canada
e-mail: ddlichti@ucalgary.ca

J. Böhm
Department of Civil, Environmental and Geomatic Engineering, University College London, London, UK
e-mail: j.boehm@ucl.ac.uk

F. Remondino and D. Stoppa (eds.), *TOF Range-Imaging Cameras*,
DOI: 10.1007/978-3-642-27523-4_6, © Springer-Verlag Berlin Heidelberg 2013

exploited to extend the model describing the relation of the quantities of interest, i.e. the point coordinates, and the raw measurements. This calibration approach is called data-driven, and needs to be distinguished from those that aim to modify the camera physically.

As range cameras have a field of view in the order of 45° by 45°, one frame is often not sufficient to record an entire object. This holds true for building interiors, cultural heritage artefacts, statues, urban objects (city furniture, trees, etc.), and many other objects. Thus frames from different positions and with different angular attitude need to be recorded in order to cover the whole object. This task is called registration and usually requires finding the orientation of the camera.

In the following sections, the basic models used in data acquisition and orientation will first be presented. Next we will concentrate on error sources and split them into random and systematic. In that section also the mitigation of errors will be discussed. The topic of the following section will be orientation of 3D camera frames. In the last section the calibration of range camera data is presented.

Range imaging is a field of technology that is developing rapidly. We concentrate on TOF cameras here, i.e., cameras that measure range directly and not by triangulation. Furthermore, we restrict ourselves to commercially available cameras that can reliably provide range information over the entire field of view.

2 Geometric Models

Photogrammetry is the scientific discipline concerned with the reconstruction of real-world objects from imagery of those objects. It is natural to extend the geometric modelling approaches developed and adopted by photogrammetrists for passive cameras to TOF range cameras. The well-accepted basis for that modelling is the pinhole camera model in which the compound lens is replaced (mathematically) by the point of intersection of the collected bundle of rays, the perspective centre (PC). The collinearity condition,

$$\mathbf{r}_i = \mathbf{r}_j^c + \lambda_{ij} \mathbf{R}_j^T \mathbf{p}_{ij} \tag{1}$$

that a point on the object of interest, $\mathbf{r}_i = \begin{pmatrix} X & Y & Z \end{pmatrix}_i^T$, its homologous point in the positive image plane, $\mathbf{p}_{ij} = \begin{pmatrix} x_{ij} - x_{p_j} & y_{ij} - y_{p_j} & -c_j \end{pmatrix}^T$, and the camera's perspective centre, $\mathbf{r}_j^c = \begin{pmatrix} X^c & Y^c & Z^c \end{pmatrix}_j^T$, lie on a straight line holds true if the incoming light rays are undistorted (Fig. 1).

Two basic sets of parameters are needed to model the central perspective imaging geometry. The first is the exterior orientation (extrinsic) parameter (EOP) set that models the camera pose, more specifically the position and angular orientation of the image space relative to object space. The EOP set thus comprises the three-element position vector of the PC, $\mathbf{r}_j^c$, and three independent angular parameters, often Euler angles $(\omega_j, \phi_j, \kappa_j)$. However parameterized, the angular elements are encapsulated in a 3×3 rotation matrix, e.g.

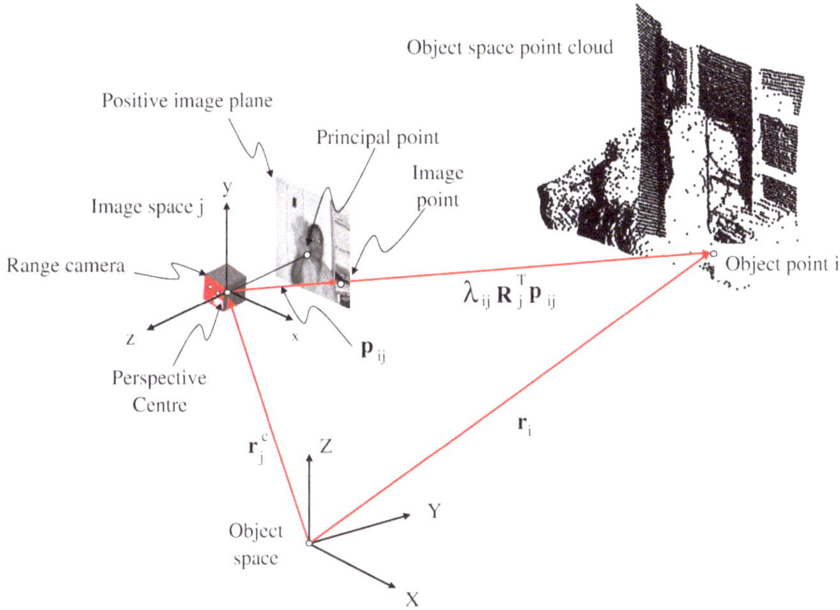

Fig. 1 TOF range camera geometric model

$$\mathbf{R}_j = \mathbf{R}_3(\kappa_j)\mathbf{R}_2(\phi_j)\mathbf{R}_1(\omega_j) \tag{2}$$

The determination of the EOPs for a TOF range camera is the subject of Sect. 3.

The second is the interior orientation (intrinsic) parameter (IOP) set that models the basic geometry inside the camera and comprises three elements. The first two are the co-ordinates of the principal point, (x_{p_j}, y_{p_j}), the point of intersection of the normal to the image plane passing through the perspective centre. The third is the principal distance, c_j, the orthogonal distance between the image plane and the PC, which is not necessarily equal to the focal length.

In passive-camera photogrammetry the unique scale factor λ_{ij} is unknown and, as a result, 3D co-ordinates cannot be estimated from a single view without additional constraints. A TOF range camera allows extension of the collinearity model with the condition that the length of the line between the PC and an object point is equal to the measured range. Thus 3D object space co-ordinates can be uniquely determined from a single range camera view

$$\mathbf{r}_i = \mathbf{r}_j^c + \frac{\rho_{ij}}{\left\|\mathbf{p}_{ij}\right\|}\mathbf{R}^T\mathbf{p}_{ij} \tag{3}$$

Following the usual convention in photogrammetry that is well suited to Gauss-Markov model formulation and least-squares estimation techniques, the extended collinearity condition can be recast from the direct form of Eq. 3 into observation equations of image point location, (x_{ij}, y_{ij})

$$x_{ij} + \varepsilon_{x_{ij}} = x_{p_j} - c_j \frac{U_{ij}}{W_{ij}} + \Delta x_{ij} \tag{4}$$

$$y_{ij} + \varepsilon_{y_{ij}} = y_{p_j} - c_j \frac{V_{ij}}{W_{ij}} + \Delta y_{ij} \tag{5}$$

where

$$(U \quad V \quad W)_{ij^T} = \mathbf{R}_j(r_i - \mathbf{r}_j^c) \tag{6}$$

and range, ρ_{ij}

$$\rho_{ij} + \varepsilon_{\rho_{ij}} = \left\| \mathbf{r}_i - \mathbf{r}_j^c \right\| + \Delta \rho_{ij} \tag{7}$$

These equations have been augmented with systematic error terms $\left(\Delta x_{ij}, \Delta y_{ij}, \Delta \rho_{ij}\right)$ and random error terms $\left(\varepsilon_{x_{ij}}, \varepsilon_{y_{ij}}, \varepsilon_{\rho_{ij}}\right)$ that account for imperfections in the imaging system, which are described in Sect. 4.

3 Orientation

Commonly several separate stations are required to entirely capture the geometry of a scene with a range camera. This might be necessary because of the limited field of view of the sensor as mentioned, its limited range, the extent of the object, self-occlusion of the object or occlusions caused by other objects. Before the data can be passed down the processing pipeline to successive steps, such as meshing and modelling, the alignment of the range measurements into a common reference frame has to be performed, often referred to as registration. Mathematically we seek the rigid body transformation defining the six parameters of translation and rotation which transforms points from the sensor coordinate system to the common or global coordinate system. As we assume the camera to be calibrated with the procedures described in Sect. 5, the scale parameter is known. Since the common coordinate system for a range camera is seldom a geodetic coordinate system, we do not specifically consider the issue of georeferencing.

There exist various approaches to solve the problem of orientation. The approaches differ in their prerequisites on the scene, prior knowledge, extra sensor measurements and the level of automation and robustness. The approaches can be categorized into (1) marker-based approaches, which require the placement of markers (e.g. planar targets, spheres, etc.) in the scene, either as control or tie points, (2) sensor-based approaches, which require additional sensors (e.g. IMU, external trackers, etc.) to be attached to the scanner in order to directly determine the position and orientation of the sensor and (3) data-driven approaches, which use the geometry and other properties of the acquired data to determine the transformations in-between scans.

The orientation of range measurements acquired with a 3D camera shares many properties with orientation of other range data, e.g. acquired with terrestrial laser scanners. Therefore many of the known solutions [1] can be applied to the orientation of 3D cameras. One property that makes 3D cameras distinctly different is the higher frame rate which offers additional processing possibilities.

3.1 Marker-Based Orientation

The marker-based registration is expected to achieve the highest accuracy. It utilises artificial targets which must be inserted into the scene. Artificial targets are well known from photogrammetry and classical surveying. Artificial targets either represent control points which are linked to some reference coordinate system or tie points. Since most TOF cameras provide both an intensity and a depth channel, both two-dimensional flat targets and three-dimensional shapes can be used as markers. The dominant shape for two-dimensional targets is white circles on a contrasting background, although checker-board patterns are also used. The dominant three-dimensional shape is spheres. Using artificial targets has the advantage of enabling measurements on 'cooperative' surfaces, i.e. surfaces of chosen reflectance properties. This removes any measurements errors due to disadvantageous material properties. Due to the limited pixel-count of many TOF cameras it can be a problem to provide markers in sufficient numbers and with sufficient size in the image domain. This problem occurs specifically in calibration. Reference [2] used infrared LEDs as markers. They are both small in image space and yet deliver precise measurements. The software tools for extracting markers from the image data can be directly adopted from close-range photogrammetric software (for two-dimensional markers) or from terrestrial laser scanning (for three-dimensional markers).

However accurate, it must be noted that marker-based approaches require extra effort for the placement and measurement of the targets. The placement of such targets may be prohibited for certain objects. For these reasons marker-less approaches are of high interest both form a practical point of view and from an algorithmic point of view.

3.2 Sensor-Based Orientation

Well-known in aerial photogrammetry, sensor-based orientation involves the integration of additional sensors to measure the pose of the camera. Typically a GNSS sensor and an inertial measurement unit (IMU) are integrated for estimation of position and orientation parameters. However as most TOF cameras are used indoors we rarely find this integration. Rather we see sensor-assisted orientation, where an IMU is used to stabilize the estimation of the orientation parameters. This approach is common in robotics [3].

3.3 Data-Driven Orientation

Data-driven registration attempts to find the transformation parameters in-between the camera stations from the sensed point cloud and intensity data itself. Some approaches reduce the complexity of the dataset by using feature extraction. Again as we can use both intensity and range data, several feature operators are available most well-known from purely image-based approaches. Reference [4] gives a comparison of some standard feature operators on TOF camera images and reports the SIFT to provide the best results. Recent work in robotics has produced novel local feature descriptors which purely use the 3D information, such as the *Point Feature Histogram* [5] and the *Radius-based Surface Descriptor* [6].

The group of algorithms for using the full point cloud geometry is the *Iterative Closest Point* (ICP) algorithms. The ICP has originally been proposed by Besl and McKay [7] for the registration of digitized data to an idealized geometric model, a typical task in quality inspection. The ICP can also be used on a sparse cloud of feature points. In its most basic form the ICP is restricted to pair-wise registration of fully overlapping datasets. Several extensions and improvements to the original algorithm have been proposed since its original publication. An overview is given by Eggert et al. [8] (we also recommend the extended version of the article which is available online) and Rusinkiewicz and Levoy [9]. It should be noted that the ICP has progressed to be the dominant registration algorithm for multi-station point cloud alignment.

4 Error Sources

Three types of errors can be distinguished and used to characterize the behaviour of instruments including TOF cameras. These are random errors, systematic errors, and gross errors.

The random errors are independent of each other (see also [10]). This applies, on the one hand, to the errors of all pixels within a frame, but on the other hand also to the errors in each pixel through the frames. Random errors in range cameras have their cause in shot noise and dark current noise. Repeating an experiment with a TOF camera will result in slightly different ranges being recorded, which are caused by the random errors. By averaging, their influence can be reduced. One way of averaging is to perform the (complex) averaging of frames if the exterior orientation as well as the scene is stable and the warm-up period of the camera has passed. In this case, averaging is performed in the time domain. Another way of averaging, performed in the spatial domain, is the modeling of the scene using geometric primitives, for example. This requires prior knowledge of the suitability and existence of these primitives within the scene. However, a group of measurements belonging to one primitive visible in one frame will have independent random errors. By applying an optimization technique, parameters of the primitives can be estimated such that the errors between measurement and scene model are minimized. This results in a scene model of better precision than the original measurements. By increasing the number of measurements, either spatially or temporally in the averaging process, the precision improves.

However, averaging does not necessarily lead to more and more accurate values, especially because of the existence of systematic errors. Those errors may stay constant during the repetition of an experiment or they may vary slowly, e.g., because of the temperature of the chip. In first instance, quantifying these errors is of interest, because it describes how much the measurement may deviate from the "true" value even if the random errors were eliminated. However, those errors can also be modeled, which is in fact an extension of the models described in Sect. 2. Among those errors are lens distortion and range errors caused by electronic cross talk. While the causes and suitable modeling strategies for these effects are known, reproducible errors for which neither the origin nor an appropriate modelling approach are known are also encountered. In this context, the distinction will be made between the physical parameters and the empirical parameters used to extend the basic models of Sect. 2.

Finally, gross errors, also called blunders, are defined as errors which do not fit to the measurement process at all. They have to be identified with suitable mathematical approaches, typically robust methods of parameter estimation, and eliminated. Those errors will not be further discussed here, but their detection and elimination is an area of ongoing research [11].

The systematic errors described in the following are camera internal errors, errors related to the operation of the camera, or errors related to the scene structure.

- Lens distortions: camera internal errors modelled with physical parameters
- Range finder offset, range periodic error and signal propagation delay: camera internal errors modelled with physical parameters
- Range error due to position in the sensor and range errors related to the recorded amplitude: camera internal error, modelled empirically
- Internal scattering: camera internal error modelled empirically and physically, respectively, by different authors
- Fixed pattern noise: camera internal error, not modelled
- Integration time errors: related to the operation of the camera, resulting in different sets of error models (e.g. for range) for different integration times
- Camera warm up errors and temperature effects during measurement: related to the operation of the camera, quantified, not modelled
- Range wrapping errors: related to the distances in the scene, modelled physically
- Scene multi-path errors: related to the scene structure, quantified by some experiments.

Motion blur, caused by the movement of the TOF camera or the movements in the scene, is a further effect. If the range to a target within a pixel's instantaneous field of view is not constant during the acquisition of a frame, then the recorded range corresponds to an average distance. Likewise, multiple objects at different distances may be within a pixel's instantaneous field of view, which will lead to an averaged distance (the mixed pixels effect). It is not appropriate to term these errors of the measurement because the measurement itself is performed as an integral over the entire pixel. However, the recorded range does not necessarily correspond to a distance from the sensor to the target which can be found in the scene itself. Therefore, these measurements should be treated as gross errors.

4.1 Lens Distortions

Radial lens distortion, or simply distortion, is one of the five Seidel aberrations and is due to non-linear variation in magnification. The systematic effect of radial lens distortion is isotropic and is zero at the principal point. Many TOF range cameras exhibit severe (i.e. tens of pixels at the edge of the image format) negative or barrel distortion. The mathematical model for radial lens distortion, Δr_{rad}, is an odd-powered polynomial as a function of radial distance, r, the Gaussian distortion profile

$$\Delta r_{rad} = k_1 r^3 + k_2 r^5 + k_3 r^7 \tag{8}$$

where k_1, k_2, k_3 are the radial lens distortion model coefficients and

$$r = \sqrt{(x - x_p)^2 + (y - y_p)^2} \tag{9}$$

The correction terms for the image point co-ordinates are easily derived from similar triangle relationships

$$\Delta x_{rad} = (x - x_p)(k_1 r^2 + k_2 r^4 + k_3 r^6) \tag{10}$$

$$\Delta y_{rad} = (y - y_p)(k_1 r^2 + k_2 r^4 + k_3 r^6) \tag{11}$$

Often only one or two coefficients are required to accurately model the distortion.

Decentring lens distortion arises due to imperfect assembly of a compound lens in which the centres of curvature of each lens element are not collinear due to lateral and/or angular offsets. It can be caused by inaccurate alignment of the sensor relative to the lens mount, i.e., the optical axis is not orthogonal to the detector array [12]. The effect is asymmetric having both radial and tangential components. Conrady's model for decentring distortion, which is expressed in terms of the radial and tangential terms, can be recast into Cartesian components

$$\Delta x_{dec} = p_1 \left(r^2 + 2(x - x_p)^2\right) + 2p_2 (x - x_p)(y - y_p) \tag{12}$$

$$\Delta y_{dec} = p_2\left(r^2 + 2\left(y - y_p\right)^2\right) + 2p_1\left(x - x_p\right)\left(y - y_p\right) \tag{13}$$

The effect of decentring distortion is typically an order of magnitude lower than that of radial lens distortion.

4.2 Scene-Independent Range Errors Modelled Physically

Three error sources in the range measurements are discussed in this subsection: the rangefinder offset (d_0), periodic errors (d_2 to d_7) and signal propagation delay errors (e_1 and e_2). The common thread among these is that they are instrumental errors that are independent of the scene structure. This is in contrast to the scattering range error (Sect. 4.4) that is also an instrumental source but it is very strongly dependent on the nature of the imaged scene.

$$\Delta\rho = d_0 + \sum_{m=1}^{3}\left[d_{(2m)}\sin\left(\frac{2^m\pi}{U}\rho\right) + d_{(2m+1)}\cos\left(\frac{2^m\pi}{U}\rho\right)\right] + e_1\left(x - x_p\right) + e_2\left(y - y_p\right) \tag{14}$$

In other rangefinding technologies, such as tacheometric equipment, the offset parameter d_0 models the offset of the range measurement origin from the instrument's vertical axis. It can also be a lumped parameter that models internal signal propagation delays and its value may be temperature dependent [13]. In the context of a TOF range camera, the offset represents the difference between the range measurement origin and the PC of the pinhole camera model. The first approximation is that the rangefinder offset d_0 is constant, but deviations from this may be modelled with a pixel-wise look-up table or a position-dependent "surface" model.

The periodic range errors are caused by odd-harmonic multiples of the fundamental frequency contaminating the modulating envelope, which results in a slightly square waveform. The physical cause of this is the non-ideal response of the illuminating LEDs [14]. The errors have wavelengths equal to fractions of the unit length, U (half the modulation wavelength). Pattinson [15] gives the mathematical explanation for the existence of the U/4-wavelength terms. The origins of the U/4- and U-wavelength errors that have been observed experimentally are not completely clear. Some (e.g. [16]) favour non-sinusoidal bases to model the periodic errors such as B-splines or algebraic polynomials (e.g., [17]).

The e_1 and e_2 terms are the signal propagation delay errors [18], also known as the clock-skew errors [19]. They are caused by the serial readout from the detector array. Their effect is a linearly-dependent range bias, a function of the column and row location, respectively.

4.3 Range Errors Depending on Position in the Focal Plane and Range Errors Related to the Recorded Amplitude

Systematic range errors depending on the recorded amplitude and the position in the focal plane are reported in ([20–23]). While arguments are brought forward for the physical relation between recorded amplitude and a range error related to individual diodes (rise and fall time of IR diodes, [21]), the assembly of many diodes for illumination of the scene prevents physical modeling.

The TOF cameras investigated with respect to the dependence of the range error on the position in the image plane feature a notable light fall-off from the center to the image borders. As no physical cause is given for this error, the relation of the range error sources amplitude and position in image plane are not resolved.

The recorded amplitude a is a function of the distance from the sensor to the illuminated footprint on the object and the object brightness itself. Furthermore, objects which are neither perfect retro-reflectors nor isotropic scatterers feature a dependence of the remitted energy on the incidence angle. Thus, object brightness (at the wavelength of the diodes) and angle of incidence may appear to have an influence on the range error, but primarily this relates to the influence on the backscattered energy.

Reference [20] reports for the SR3000 range errors of 40 cm for low amplitudes. Additional maximum errors of 25 cm depending on the position in the image plane are shown. However, these image plane errors are concentrated strongly in the corners.

The models to describe these systematic offsets need to be developed since no physical basis exists. The functions are typically chosen to have as few parameters as possible in order to prefer a simple model. On the other hand, the remaining errors, after subtraction of the modelled systematic behaviour should only be random. In [20] a hyperbolic function with parameters h_1, h_2, h_3 is chosen. This model fits to the general observations that range errors are positive and large for low amplitudes, rapidly become smaller for larger amplitudes and do not change much for higher amplitudes. The equation for relating the range error to the observed amplitude is:

$$\Delta\rho = -\frac{h_2}{h_3}a + \sqrt{\frac{h_2^2}{h_3^2}a^2 - 2\frac{h_1}{h_3}a + \frac{1}{h_3}} \tag{15}$$

It has to be noted that these descriptions should only be used for one range camera, as no physical principle applicable to all range cameras builds the basis. See Fig. 2 for an example.

4.4 Internal Scattering

The echo of the emitted optical signal is, in the geometrical model of the camera, focused by the lens onto the individual pixels. In reality, however, the signal focused

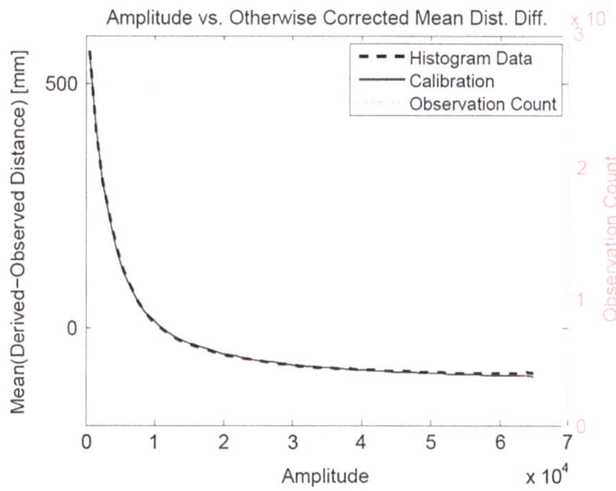

Fig. 2 Systematic range errors as function of the observed amplitude. The dashed line represents the differences between the observed and the reference distance. The solid line is the model that describes the systematic range error [20]

onto one pixel is scattered partially within the camera and distributed over the sensor area. This is caused by multiple reflections within the camera: between the lens, the optical filter, and the sensor. Therefore, the signal observed at an individual pixel is a mixture of the focused light, returned from the geometrically corresponding pixel footprint on the object, and the scattered light reflected at other pixels and thus corresponding to other parts of the object. This is illustrated in Fig. 3. Three targets at different distances, and therefore with different phase angles, are imaged by the range camera. Because target 1 is a strong reflector and close to the sensor, a notable portion of the focused light is scattered and influences the focused light backscattered from targets 2 and 3. Because target 3 in the given example features a low amplitude due to the larger distance, the scattered light from target 1 will influence the derived range stronger than for the "brighter" target 2. The impact on the observed amplitudes may be small, but phase angle measurements and derived distances are strongly affected in images with high amplitude and depth contrast. Such high contrasts are typical for systems with active illumination.

Scattering can be described by the convolution of a point spread function (PSF) with the "unscattered" image. This assumes that scattering in TOF cameras can be described as a linear phenomenon, which is – experimentally – verified by different studies [25]).

[26] Conducted experiments with an SR3000 and images with a high contrast in depth (0.73 to 1.46 m) and target reflectivity (retro-reflective foil). Scenes with and without the target in the foreground were subtracted in the complex domain. Maximum distortions in the distance were found to be 40 cm. Reference [27] performed experiments with an SR3000 and an SR4000 using two planar objects. The foreground object covered half the scene. The scattered signal from the foreground object to the pixel which images (geometrically) the background resulted in "errors" as large as the distance between the two objects. For the SR4000 scattering was more than one order of magnitude smaller. Differently, [28] could not assert scattering effects in their experiments.

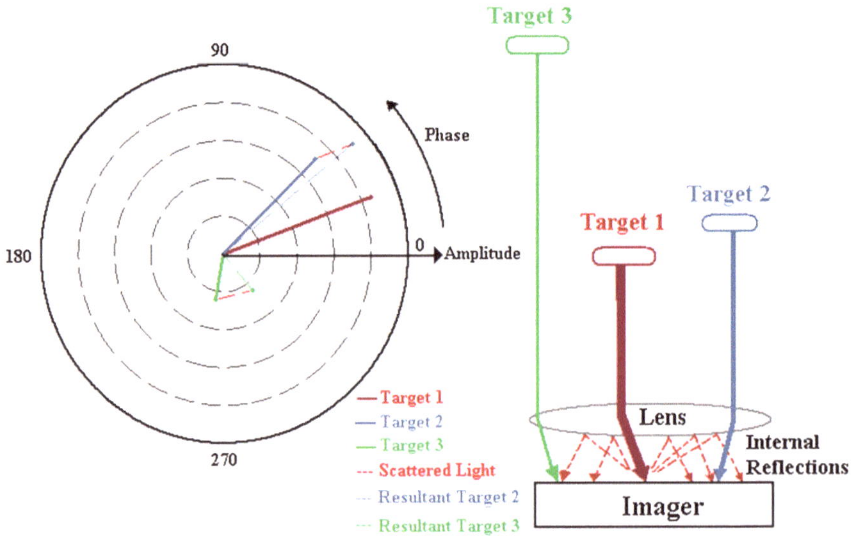

Fig. 3 Targets at different distances and having different amplitudes of the geometrically recorded signal. *Left*: the amplitude and phase angle are shown as a complex number. *Right*: scattering from target 1 to targets 2 and 3 is illustrated [24]

4.5 Fixed Pattern Noise

Each pixel can be considered to be individually the cause of a systematic range error. This is called fixed pattern noise. Piatti [29] attributes it to the "imperfect manufacturing process and different material properties in each sensor gate, which yields a pixel-individual fixed measurement offset". In [30] it is argued that modeling this error for the PMD-camera does not lead to an increase in precision.

4.6 Errors Dependent on Integration Time

In the focal plane of the TOF camera the backscatter of the emitted light is collected. In order to have a high signal-to-noise ratio (SNR) for precise measurement at each pixel, the measurement period should be as long as possible. On the one hand, this can lead to saturation effects, and on the other hand, moving objects in the scene or movement of the camera may motivate a short integration time in order to avoid motion blur effects. Thus, the integration time can typically be set by the user and is adjusted to the observed scene. The SR4000 allows, e.g., setting the integration time approximately between 0.3 and 25 μs, whereas the PMD CamCube bounds it approximately by 0.01 and 50 μs.

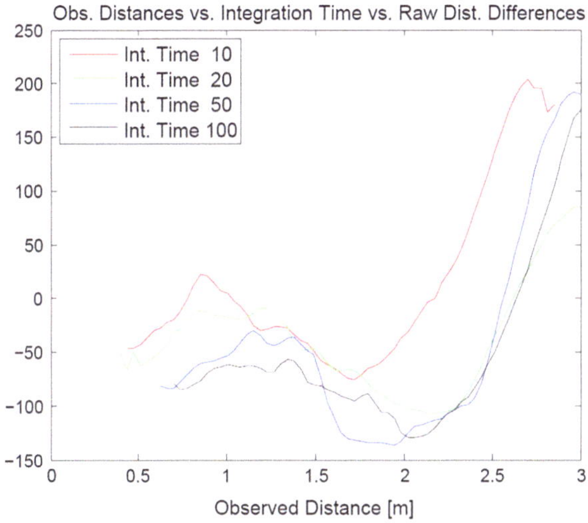

Fig. 4 Observed distances vs. range error for different integration times of the SR3000, between 0.4 and 200 μs. The periodic range measurement errors (see 6.4.2) are clearly visible, but also that the integration time has an impact on the phase of the periodic errors. The deviations from a harmonic originates in other error sources (recorded intensity), which also varied in the acquired data. The period of these range errors is 1.75 m, which is an eighth of the modulation wavelength of 15 m. [20]

Different authors have reported that the systematic errors described above depend on the integration time selected by the user ([20], [23]). The influence on the range errors of the chosen integration time is shown in Fig. 4.

4.7 Camera Warm Up Errors

Controlled tests conducted by [23, 31] show that range measurements between a stationary camera and a stationary target exhibit a significant transient error response as a function of time after the camera is powered up due to strong temperature dependence. The magnitude of this drift in the older SR3000 model of the SwissRanger camera is reported to be on the order of several centimetres [23]. They also show that this effect can be reduced to the centimetre level by introducing an optical reference into the camera. The known internal path of light passing through an optical fibre allows correction of measured ranges for the temperature-caused drifts. The warm-up transient effect in the newer SR4000 is reported by [31] to be smaller, i.e. on the order of several millimetres, but takes tens of minutes to decay. They suggest that a warm-up period of 40 min be observed prior to camera use. An example of the warm-up effect is given in Fig. 5.

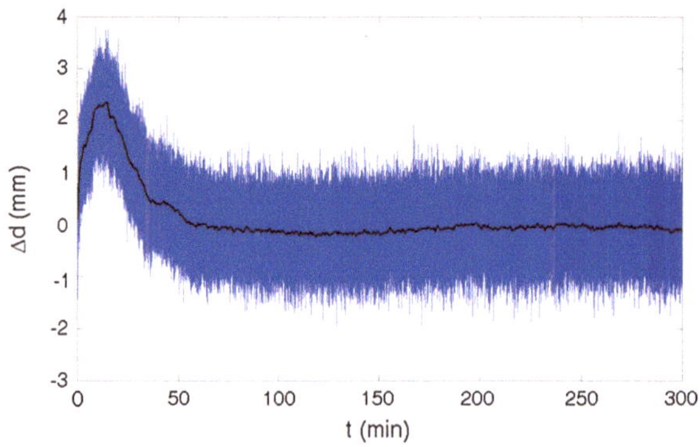

Fig. 5 Range error, Δd, due to the camera warm-up effects in SR4000 data collected every 30 ms for 5 h. A diffusely-reflecting, planar target was imaged at normal incidence from a range of 2.1 m. The range of the 60 s moving-average trend is 2.6 mm. In this example the transient dies out after about 60 min but the long-term stability thereafter is very good

The temperature of the camera does not only vary when the camera is switched on, but also depends on the work load of the camera which is shown in Fig. 6, [15]. Three distinct phases are depicted and separated by vertical bars: a first phase of low frame rate, then a phase of high frame rate (up to 10 fps) and then a phase of low frame rate again. The distance which is measured to fixed targets varies for several centimeters depending on the frame rate (and consequently the temperature of the camera) at which the camera is driven.

4.8 Ambiguity Interval

Ranges determined by the phase-difference method are inherently ambiguous since the phase measurements are restricted to [0, 2π). For a single-tone system, the maximum unambiguous range or ambiguity interval, U, is determined by the modulation frequency, f_{mod},

$$U = \frac{c}{2f_{mod}} \tag{16}$$

For the SR3000 the nominal modulation frequency and maximum unambiguous range are 20 MHz and 7.5 m, respectively. The SR4000 features the ability to set the frequency to one of several pre-defined values; the default is 30 MHz, for which U is 5 m.

The range to a target beyond the ambiguity interval is mapped onto [0, U), resulting in a discontinuous wrapped range image or wrapped phase map (Fig. 7). Phase unwrapping must be performed to estimate the integer ambiguity—the number of whole cycles between the camera and target—at each pixel location, thereby removing the discontinuities. Jutzi [32] has demonstrated that it is possible

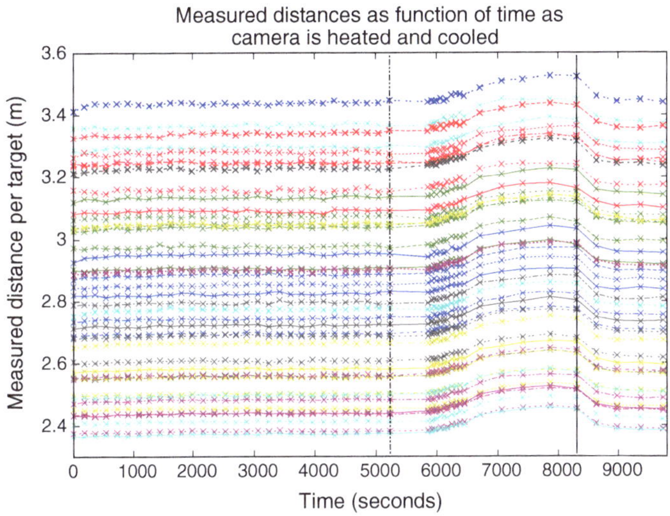

Fig. 6 Range measurements to multiple targets at varying frame rate [15]

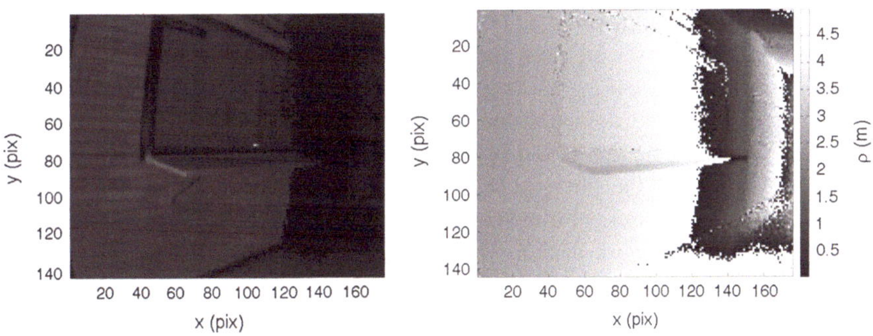

Fig. 7 *Left*: SR4000 amplitude image captured in a hallway. *Right*: corresponding wrapped range image showing two discontinuities

to unwrap the phase of a range camera image with unwrapping 2D algorithms. He suggests making use of the measurement-confidence value available with some cameras' output to guide the unwrapping.

4.9 Multi-Path Effects in Object Space

Similar to the scattering of the signal inside the camera (Sect. 4.4), parts of the emitted signal may be scattered in object space by closer objects, casting the signal further to other objects, and from there to the corresponding pixel in the focal plane. Thus, the footprint of the pixel on the object is illuminated twice: once directly, and once via a longer object multi-path. Again, the relative strength of the multi-path signal to the direct illumination signal determines the size of the range error. The final range and amplitude is found by the complex addition of all direct and multi-path signals incident onto each pixel. Descriptions of this phenomenon are found in [33, 34].

5 System Calibration

5.1 Purpose of Calibration

The geometric positioning models given by Eqs. 4, 5 and 7 are mathematical simplifications of the imaging process. In reality the many influencing variables described in Sect. 4 cause perturbations from the idealized geometric conditions, which if ignored can degrade the accuracy of a captured 3D point cloud. In calibration one seeks to estimate the coefficients of the models for the instrumental systematic errors. In doing so the influence of other error sources, such as the ambient atmospheric conditions and scene-dependent artefacts like multi-path and scattering, must be eliminated as best as possible to prevent biases. Currently the estimable parameter set includes the lens distortions, rangefinder offset, periodic range errors, clock skew errors as well as amplitude-dependent range errors.

Förstner [35] breaks down the camera calibration (and orientation) problem into four central tasks. The first is sensor modelling, treated in Sects. 2 and 4, in which an understanding of the physics of image formation, possibly guided by model identification procedures, is transformed into the mathematical models. The second is geometric network design where the goal is to maximize the accuracy of coefficient estimation through the judicious choice of a configuration of sensor and calibration primitive locations. This subject is touched upon briefly in this section; greater detail can be found in the cited references. The third and fourth tasks are error handling (i.e. procedures for outlier identification) and automation, respectively, neither of which is treated here. However, there is ongoing research to address these problems also in neighbouring disciplines [11].

5.2 Calibration Approaches

Two basic approaches to TOF range camera calibration can be considered. The first is laboratory calibration, the name of which implies that it is conducted in a controlled setting. Specialized facilities (e.g. a geodetic calibration track) are used to very accurately determine (usually) a subset of calibration model parameters.

Multiple calibration procedures (e.g. range error calibration over a baseline and lens distortion calibration over a target field) are generally required for complete system characterization. The principal advantage of this approach is that the network design tightly is controlled by the specialized observation conditions, so no parameter correlation problems are explicitly encountered.

The rather restrictive requirement for special facilities has been a driving force behind the development of self-calibration methods. Though several variants (described below) exist for TOF range cameras, they share a common underlying premise: a strong network of geometric primitives (structured point targets and/or planar surfaces) is imaged and all model variables (the IOPs augmented with systematic error model terms, the EOPs and the object primitive parameters) are simultaneously estimated from the redundant set of observations according to some optimality criterion (e.g. the weighted least-squares principle). It is a very flexible approach in that all information sources (system observations and external constraints) can be incorporated and a holistic method in which both individual component errors and system assembly errors are included in the models. Absolute knowledge of the object space primitive parameters is not required, though it can be readily included in the solution. Since the facility requirements are minimal, self-calibration may be performed in a laboratory for pre- (or post-) calibration or at a job site if the stability of the camera's interior geometry is in question.

5.3 Self-Calibration Methods

The available self-calibration methods are first categorized as being range-camera-only methods or joint-setup (with a passive camera) methods. Three variants of the former are first described: the two-step independent method; the two-step dependent method; and the one-step integrated method. Data acquisition for either category may comprise still images or video sequences, which allow for greater random error mitigation via image averaging and greater redundancy.

In the two-step independent method, the camera-lens and range-error calibrations are performed as separate processes using separate facilities. First, an established procedure is used for the camera-lens calibration from x and y observations of targets in a network of convergent images [36]. Convergent imaging is needed to lower functional dependencies between the EOPs and IOPs; see [37]. Then, a planar target is imaged at normal incidence to determine the range-error parameters. Kahlmann and Ingensand [23] use a small, planar target moved along an interferometric calibration track whereas [31] use an extended planar target positioned with parallel tape measures. The orientation can also be performed by space resection of the camera from independently-surveyed targets on the plane. Regardless of the orientation method used, reference ranges between each image's PC and the target plane, described by the unit surface normal vector **n** and distance parameter d, are computed as follows

$$\rho^{\mathrm{ref}} = \frac{\mathrm{d} - \mathbf{n}^{\mathrm{T}}\mathbf{r}_{\mathrm{j}}^{\mathrm{c}}}{\mathbf{n}^{\mathrm{T}}\mathbf{R}_{\mathrm{j}}^{\mathrm{T}}\mathbf{p}_{\mathrm{ij}}} \left\| \mathbf{p}_{\mathrm{ij}} \right\| \tag{17}$$

The reference ranges are compared with the observed ranges to derive a dense set of range differences from which the range-error parameters are estimated by least-squares.

A common facility is used for both calibration processes in the two-step dependent method. The camera-lens calibration is first performed with an established procedure using the x and y observations of targets on a planar surface observed in a network of both convergent and orthogonal images. The camera-plane orientation is established by the camera calibration so there is no need for an independent orientation means. The reference ranges can then be computed from the orthogonal camera stations to points on the plane [38, 39] or to the target centers [40] and used for the range-error calibration as described above.

The third approach is the one-step integrated method in which both sets of calibration parameters (camera-lens and range-error) are estimated simultaneously [27]. A planar target field is imaged from both convergent and orthogonal camera locations. To prevent scattering errors from biasing the solution, the range observations from the convergent stations are excluded from the bundle adjustment. In this approach the camera orientation is performed concurrently and there is no explicit computation of reference ranges.

In the joint-setup methods a high-resolution, pre-calibrated passive digital camera is used to aid the range camera calibration. The two cameras are rigidly mounted together in a frame such that there is a high degree of overlap of their fields-of-view and their relative orientation is invariant. Reference [16] presents a two-step (i.e. camera-lens calibration followed by range calibration) joint-setup method while [17] proposes an integrated joint-setup approach that incorporates both convergent and normal images in the network.

The principal advantage of the range-camera-only methods is that no auxiliary equipment is required to perform the self-calibration. Though in principle a more rigorous method, the one-step integrated approach suffers from high correlation between the rangefinder offset and the PC position brought about by the use of normal-incidence imaging of the planar target field. This correlation exists in the two-step methods as well, just not explicitly. However [41] have demonstrated that the precision and accuracy differences between the three range-camera-only methods were not of any practical significance. The advantage of the joint-setup approach is the decoupling of the IOPs and EOPs parameters. Furthermore, it is a logical procedure to pursue if one wishes to colourize point clouds captured with the range camera with colour imagery from the passive digital camera.

5.4 *Model Formulation and Solution*

Regardless of the exact self-calibration procedure, the first step in the numerical model solution is formulation of the deterministic part of the linearized Gauss-Markov model:

$$\mathbf{Ax} = \mathbf{b} + \mathbf{v} \tag{18}$$

where $\mathbf{x}$ denotes the vector of unknown parameters; $\mathbf{b}$ is the vector of observations; $\mathbf{A}$ is the Jacobian of the observations with respect to the parameters; and $\mathbf{v}$ is the vector of error estimates (residuals). If, for example, one considers the point-based, one-step self-calibration procedure, then this system can be partitioned row-wise according to observation group (x,y,ρ) and column-wise according to parameter group (e: EOPs; i: IOPs; o: object point co-ordinates)

$$\begin{pmatrix} \mathbf{A}_{\mathrm{xe}} & \mathbf{A}_{\mathrm{xi}} & \mathbf{A}_{\mathrm{xo}} \\ \mathbf{A}_{\mathrm{ye}} & \mathbf{A}_{\mathrm{yi}} & \mathbf{A}_{\mathrm{yo}} \\ \mathbf{A}_{\mathrm{\rho e}} & \mathbf{A}_{\mathrm{\rho i}} & \mathbf{A}_{\mathrm{\rho o}} \end{pmatrix} \begin{pmatrix} \mathbf{x}_{\mathrm{e}} \\ \mathbf{x}_{\mathrm{i}} \\ \mathbf{x}_{\mathrm{o}} \end{pmatrix} = \begin{pmatrix} \mathbf{b}_{\mathrm{x}} \\ \mathbf{b}_{\mathrm{y}} \\ \mathbf{b}_{\mathrm{\rho}} \end{pmatrix} + \begin{pmatrix} \mathbf{v}_{\mathrm{x}} \\ \mathbf{v}_{\mathrm{y}} \\ \mathbf{v}_{\mathrm{\rho}} \end{pmatrix} \tag{19}$$

The stochastic model is defined by

$$\mathrm{E}\{\mathbf{v}\} = \mathbf{0} \tag{20}$$

and

$$\mathrm{E}\{\mathbf{v}\mathbf{v}^{\mathrm{T}}\} = \mathbf{C} \tag{21}$$

where $\mathbf{C}$ is symmetric, positive-definite and diagonal if uncorrelated observational errors are assumed.

The system of Eq. (19) must be subjected to a set of minimum object space datum constraints such as the inner constraints, represented by the design matrix $\mathbf{G}$, imposed on object points only (e.g. [42]).

$$\mathbf{G}_o^T \mathbf{x}_o = \mathbf{0} \tag{22}$$

The least-squares solution of this system of equations is performed iteratively in order to obtain optimal parameter estimates. Their covariance matrix quantifying parameter solution quality is obtained directly from the solution. Further details about the least-squares solution and quality measures can be found in Kuang [43], for example.

6 Summary

This chapter summarized approaches to estimate the orientation of TOF cameras and model the imaging process. The geometric principles of the imaging process are based on the collinearity of the object point, the camera perspective centre, and the image point. This is augmented by the direct range observation performed by TOF cameras, providing the distance from the perspective centre to the object point.

However, systematic deviations from this model exist. Thus, range cameras need to be calibrated. Depending on the range camera used and the experimental design, errors may be larger than 10 cm. If possible, the physical cause for these systematic errors should be found and modelled. This was shown for periodic range errors. In other cases, an empirical modelling approach must be chosen, as the cause for reproducible systematic errors is not known or too complicated for modelling (e.g. error related to amplitude). Other causes of error are range wrap and scattering. Range wrapping can be corrected with weak assumptions on the scene. Scattering, however, needs to be treated differently. Its removal by deconvolution techniques is the first step in processing range data, i.e. before orientation and calibration.

Orientation and calibration are not independent of each other. If the stability of a range camera does not allow determining calibration parameters once for a longer time period (e.g. one year), self-calibration by exploitation of project data is necessary. In such a case, the orientation and calibration are solved simultaneously.

With the on-going development of TOF camera technology, the set of systematic errors becomes smaller and smaller. Still calibration remains important because it is the appropriate means of quantifying both random and systematic errors.

The on-going development with respect to TOF camera resolution and reduced noise in range and amplitude observation will also increase the accuracy of estimating the orientation.

References

1. N. Pfeifer, J. Boehm, Early stages of LiDAR data processing, *Advances in Photogrammetry, Remote Sensing and Spatial Information Sciences*, CRC Press (2008)
2. T. Kahlmann, F. Remondino, H. Ingensand, Calibration for increased accuracy of the range imaging camera SwissRanger. Int. Arch. Photogram. Remote. Sens. Spat. Inf. Sci. Vol. XXXVI, part 5, 136–141 (2006)
3. D. Dröschel, S. May, D. Holz, P. Plöger and S. Behnke, Robust Ego-Motion Estimation with TOF Cameras, Proceedings of the 4th European Conference on Mobile Robots (ECMR), Dubrovnik, Croatia, Sept 2009
4. S. May, D. Droeschel, D. Holz, C. Wiesen and S. Fuchs, 3D Pose Estimation and Mapping with Time-Of-Flight Cameras, IEEE/RSJ International Conference on Intelligent Robots and Systems (IROS), Workshop on 3D Mapping, Nice, France, 2008
5. R. B. Rusu, N. Blodow, M. Beetz, Fast Point Feature Histograms (FPFH) for 3D registration, *2009. ICRA '09. IEEE International Conference on Robotics and Automation*, pp.3212–3217 (2009)
6. Z. Marton, D. Pangercic, N. Blodow, J. Kleinehellefort, M. Beetz, General 3D modelling of novel objects from a single view, *2010 IEEE/RSJ International Conference on , Intelligent Robots and Systems (IROS)*, pp. 3700–3705, 18–22 Oct 2010

7. P.J. Besl, N.D. McKay, A method for registration of 3-D shapes. IEEE Trans. Pattern Anal. Mach. Intell. **14**(2), 239–256 (1992)
8. D.W. Eggert, A.W. Fitzgibbon, R.B. Fisher, Simultaneous Registration of Multiple Range Views for Use in Reverse Engineering of CAD Models. Comput. Vis. Image Underst. **69**(3), 253–272 (1998)
9. S. Rusinkiewicz and M. Levoy, Efficient variants of the ICP algorithm, *Proceedings of the Third Intl. Conf. on 3D Digital Imaging and Modeling*, pp. 142–152, 2001
10. ISO/IEC Guide 98-1, *Uncertainty of measurement – Part 1*: *Introduction to the expression of uncertainty in measurement* (Geneva, Switzerland, 2009)
11. M. Reynolds, J. Doboš, L. Peel, T. Weyrich, G.J. Brostow, Capturing Time-Of-Flight Data with Confidence, *CVPR* (2011)
12. D.C. Brown, Decentering distortion of lenses. Photogramm. Eng. **32**(3), 444–462 (1966)
13. J.M. Rüeger, *Electronic Distance Measurement: an Introduction, 3rd edtion* (Springer-Verlag, Heidelberg, 1990)
14. H. Rapp, M. Frank, F.A. Hamprecht, B. Jähne, A theoretical and experimental investigation of the systematic errors and statistical uncertainties of Time-Of-Flight-cameras. Int. J. Intell. Syst. Technol. Appl. **5**, 402–413 (2008)
15. T. Pattinson, *Quantification and Description of Distance Measurement Errors of a Time-Of-Flight Camera*, MSc thesis (University of Stuttgart. 2010)
16. M. Lindner, I. Schiller, A. Kolb, R. Koch, Time-Of-Flight sensor calibration for accurate range sensing. Comput. Vis. Image Underst. **114**(12), 1318–1328 (2010)
17. M. Shahbazi, S. Homayouni. M. Saadatseresht and M. Satari, Range camera self-calibration based on integrated bundle adjustment via joint setup with a 2D digital camera. *Sensors*, 11 (9), 8721–8740 (2011)
18. S. Fuchs and S. May, Calibration and registration for precise surface reconstruction with TOF cameras. Proceedings of the dynamic 3D imaging workshop (Heidelberg, 11 Sept 2007) 9 pp
19. H. Du, T. Oggier, F. Lustenburger and E. Charbon, A virtual keyboard based on true-3D optical ranging, Proceedings of the British Machine Vision Conference (Oxford, 5–8 September 2005) p 10
20. W. Karel and N. Pfeifer. Range camera calibration based on image sequences and dense, comprehensive error statistics. In: Beraldin, J.-A., Cheok, G.S., McCarthy, M., Neuschaefer-Rube, U. (Eds.). 3D Imaging Metrology. In: Proceedings of SPIE Vol 7239. 19–20 Jan, pp. 72390D1–72390D12, 2009
21. S. Fuchs and G. Hirzinger. Extrinsic and depth calibration of TOF-cameras. In: Proc. IEEE Conference on computer vision and pattern recognition, CVPR, Anchorage, 24–26 June, 6 p. (2008)
22. Lindner, M., Kolb, A.,. Calibration of the intensity-related distance error of the PMD TOF-camera. In: Intelligent robots and computer vision XXV: Algorithms, techniques, and active vision. In: Casasent, D.P., Hall, E.L., Röning, J. (Eds.), Proc. of the SPIE. Boston, MA, 9 September, 8 p (2007)
23. T. Kahlmann, H. Ingensand, Calibration and development for increased accuracy of 3D range imaging cameras. J.Appl. Geodesy **2**(1), 1–11 (2008)
24. W. Karel, S. Ghuffar, N. Pfeifer, Quantifying the distortion of distance observations caused by scattering in Time-Of-Flight range cameras. Int. Arch. Photogram. Remote Sens Spat Inf. Sci. **38** (Part 5), 316–321 (2010)
25. T. Kavli, T. Kirkhus, J.T. Thielemann and B. Jagielski. Modelling and compensating measurement errors caused by scattering in Time-Of-Flight cameras. In: Huang, P.D., Yoshizawa, T., Harding, K.G. (Eds.). Two- and three-dimensional methods for inspection and metrology VI. In: Proceedings of SPIE Vol 7066. 28 Aug, pp. 706604-1–10, 2008
26. W. Karel, S. Ghuffar, N. Pfeifer, Modelling and compensating internal light scattering in time of flight range cameras. Photogramm. Rec. **27**(138), 155–174 (2012)
27. D.D. Lichti, C. Kim, S. Jamtsho, An integrated bundle adjustment approach to range-camera geometric self-calibration. ISPRS J. Photogram. Remote Sens. **65**(4), 360–368 (2010)

28. F. Chiabrando, D. Piatti, F. Remondino, SR-4000 TOF camera: further experimental tests and applications to metric surveys. Int. Arch. Photogram. Remote. Sens. Spat. Inf. Sci. **XXXVIII, Part 5**, 149–154 (2010)
29. D. Piatti. Time of flight cameras: tests, calibration and multi-frame registration for automatic 3D object reconstruction. Ph.D. thesis, Politecnico di Torino, Italy. (2011)
30. M. Lindner, A. Kolb, Lateral and Depth Calibration of PMD-Distance Sensors. Adv. Visual. Comput. Springer **2**, 524–533 (2006)
31. F. Chiabrando, R. Chiabrando, D. Piatti, F. Rinaudo, Sensors for 3D imaging: metric evaluation and calibration of a CCD/CMOS Time-Of-Flight camera. Sensors **9**(12), 10080–10096 (2009)
32. B Jutzi., Investigations on ambiguity unwrapping of range images. *Int. Arch.Photogram. Remote Sens. Spat. Inf. Sci.* 38 (Part 3/W8), 265–270, 2009
33. D. Falie and V. Buzuloiu, Further investigations on TOF cameras distance errors and their corrections. European Conference on Circuits and Systems for Communications, 197–200 (2008)
34. A.A. Dorrington, J.P. Godbaz, M.J. Cree, A.D. Payne, L.V. Streeter, Separating true range measurements from multi-path and scattering interference in commercial range cameras. Proceedings of SPIE 7864 (2011)
35. W. Förstner, Generic estimation procedures for orientation with minimum and redundant information. In: Gruen, A., and Huang T.S. (Eds.). *Calibration and Orientation of Cameras in Computer Vision*, 63–94 (2001)
36. C.S. Fraser, Photogrammetric camera component calibration: a review of analytical techniques. In: A. Gruen and T.S. Huang (Eds.). *Calibration and Orientation of Cameras in Computer Vision*, 95–121 (2001)
37. J. F. Kenefick, M. S. Gyer and B. F. Harp, Analytical self-calibration. Photogram. Eng., 38 (11), 1117–1126 (1972)
38. S. Robbins, B. Murawski and B. Schroeder, Photogrammetric calibration and colorization of the SwissRanger SR-3100 3-D range imaging sensor, Opt. Eng. 48 (5), 053603-1–8 (2009)
39. W. Karel, Integrated range camera calibration using image sequences from hand-held operation. Int. Arch. Photogram. Remote Sens Spat Inf. Sci. **37** (Part B5), 945–951 (2008)
40. J. Boehm, and T. Pattinson, Accuracy of exterior orientation for a range camera. Int. Arch. Photogram. Remote. Sens. Spat. Inf. Sci. 38 (Part 5) [On CD-ROM] (2010)
41. D.D. Lichti, C. Kim, A comparison of three geometric self-calibration methods for range cameras. Remote Sensing **3**(5), 1014–1028 (2011)
42. C.S. Fraser, Optimization of precision in close-range photogrammetry. Photogram. Eng. Remote Sens. **48**(4), 561–570 (1982)
43. S. Kuang, *Geodetic Network Analysis and Optimal Design: Concepts and Applications* (Ann Arbor Press, Chelsea, 1996)

TOF Cameras for Architectural Surveys

Filiberto Chiabrando and Fulvio Rinaudo

1 Introduction

Digital photogrammetry developments and the massive use of LiDAR technology have led to a radical change in metric survey approaches for the documentation of Cultural Heritage. In particular, a rapid change has been taken place from 2D representations, which were always considered the only way of obtaining architectural knowledge, to 3D geometric and photorealistic representations.

Before the development of fully 3D acquisition systems, and the improvement in digital photogrammetric techniques (which are more and more oriented towards the automatic image-matching algorithms and therefore towards dense point cloud acquisition), the goal of the surveyors of architectural objects was to extract the geometric shape ofthe surveyed item using traditional systems, such as total stations, levels, close-range photogrammetry or direct measurements. These techniques, and the related recorded data, were usually used to create conventional 2D drawings, such as plans, elevations and sections.

The above mentioned approach could be defined as intelligent, rational and manual since because the surveyor selected the points that were needed to describe the shape of the surveyed object during the acquisition of the primary data (angles, distances, plotting of stereo-images).

Today the scenario has changed: the new acquisition systems allow surveyors to work with point clouds that have been acquired without any understanding of the shape of the object. The user has to interpret and describe the searched shapes on the basis of this "unintelligent" geometry.

F. Chiabrando (✉) · F. Rinaudo
Architecture and Design Department, Politecnico di Torino,
Viale Mattioli 39 10125 Turin, Italy
e-mail: filiberto.chiabrando@polito.it

F. Rinaudo
e-mail: fulvio.rinaudo@polito.it

F. Remondino and D. Stoppa (eds.), *TOF Range-Imaging Cameras*,
DOI: 10.1007/978-3-642-27523-4_7,

This radical change, which started with the introduction of Terrestrial Laser Scanners (TLS) at the end of the nineties, has undergone an interesting development over the last ten years with the introduction of new TLS systems, thanks to the recent developments in image-matching techniques (digital photogrammetry [13, 14, 33, 35]) and to new instruments: e.g. Time of Flight (TOF) cameras.

Unfortunately, the data acquired or extracted from the above mentioned systems and techniques cannot be used directly to produce 2D representations, due to the difficulty involved in extracting the geometric primitives (e.g. lines, surfaces, volumes etc.) from a 3D point cloud. Moreover, several manual interventions usually have to be performed on the extracted data, since no automatic or reliable procedures are so far available. In order to represent the correct object geometries, it is necessary to measure and extract the break-lines from the 3D survey data that allow the artefact to be described at the requested representation scale.

On the contrary the new 3D acquisition systems allow surveyors to be supplied during the 3D modelling phase. The creation of these models is actually based on different techniques (Tin, Splines, Nurbs etc.) that allow the shape of the objects to be described with a very accurate metric precision.

Hence, each survey technique has pros and cons and each allows an object to be surveyed with a different accuracy; therefore, in order to achieve a more accurate and complete survey, it is usually necessary to integrate information from more than one of the aforementioned techniques.

In the following sections, TOF cameras are analysed to check their capacity to produce point clouds of the same quality as those produced by TLS and to point out their expected applications but also to show their limits, due to the current technology developments, in terms of hardware and software.

2 TOF Camera Analysis for Architectural Surveys

TOF cameras allow 3D point clouds which are almost comparable with those of traditional LiDAR instruments to be acquired at video frame rates. Using these cameras, a bundle of distances of a two-dimensional array is determined simultaneously for each pixel. The measured distances are obtained from the time that an emitted signal takes to return to the camera. A near infra-red light, modulated in amplitude at a radio frequency, is emitted by a set of integrated light emitting diodes (LEDs) to illuminate the scene. The back-scattered light is focused on a detector array and demodulated at each detector site. The acquisition of cross-correlation measurements from four successive integration periods allows the phase difference from which the range is derived, the amplitude and the offset of the received signal to be determined. The range camera outputs are usually an amplitude image and a co-located 3D point cloud, which is computed from the image of the ranges.

Two main variations of the TOF principle have been implemented above all: one measures the distance by means of direct measurement of the runtime of a travelled light pulse using arrays of single-photon avalanche diodes (SPADs) [1]. The other method uses amplitude modulated light, and obtains distance information by measuring the phase difference between a reference signal and the reflected signal [19].

While complex readout schemes and low frame rates have so far prevented the use of SPAD arrays in commercial 3D-imaging products, the latter category has already been implemented successfully in several commercially available 3D camera systems.

Only few application of TOF cameras for metric surveys have been published up to now [11, 12, 38] and all the tests were developed on small sized objects: in those papers the modeling phase was not analyzed and TOF cameras was used as LiDAR instruments without considering the different origin of the measurements.

Therefore in the following a short but exhaustive analysis of the possible systematic errors which affect TOF measurements will be described in order to allow a correct use of TOF cameras in architectural metric surveying.

The error sources of primary TOF data can be classified in four groups [24].

The first group includes random errors due to shot noise and dark noise [20].

The second group, (scene-dependent errors) comprises systematic effects due to the ambient imaging conditions (e.g. external temperature) as well as effects due to the scene structure, including mixed pixels, multi-path reflections and the scattering artifact. The latter is a bias in both the amplitude and phase of the measured signal from background targets which is caused by the internal scattering of light reflected from a bright, foreground object [29]. The phase error is realized as a range bias.

The third group includes errors due to camera operating conditions, such as the warm-up time [9] and the selected integration time. The integration time can have a direct effect on the rangefinder offset parameter [16] and, when changed, can cause short and long-period temporal effects on the range measurements [17]. These effects are managed by warming up the camera for a sufficient period prior to data acquisition, and using a constant integration time. The final group comprises scene-independent errors, such as lens distortions, rangefinder offset, range scale error, periodic errors and latency errors [22].

The aforementioned scene-dependent or scene-independent errors, which could be defined as systematic errors, need to be corrected using suitable pre-calibration procedures [15, 23, 25, 34, 41] or some post-processing solutions [9] in order to refine the measurement accuracy.

A Swiss-Ranger-4000 (SR-4000) camera was employed in the following tests and examples. This camera is characterized by a 176×144 pixel array, and a working range of 0.3–5 m (Table 1). For more details about the camera specifications see www.mesaimaging.com.

A study on the influence of distance measurements of the camera orientation, with respect to the observed object, and an investigation on the influence of object reflectivity on the camera distance measurement accuracy and precision have been

Table 1 The Swiss ranger SR-4000 TOF camera

Technical specifications SR-4000	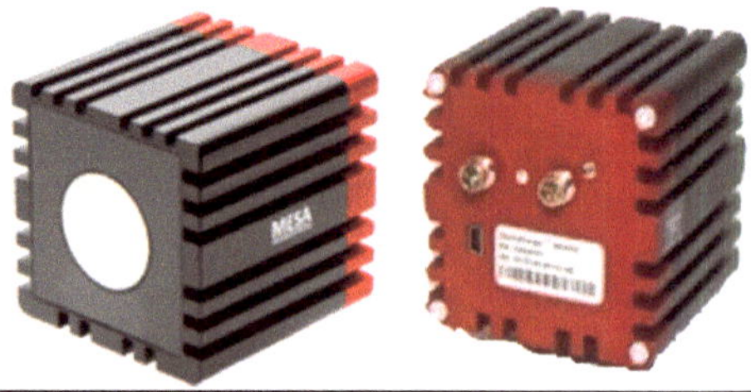
Focal length [mm]	10
Pixel array size [-]	176 (h) × 144 (v)
Pixel pitch [mm]	40
Field of view [°]	43.6 (h) × 34.6 (v)
Working range with standard settings [m]	0.3–5.0
Repeatability (1σ) [mm]	4 (typical)—7 (maximum) (@ 2 m working range and 100 % target reflectivity)
Absolute accuracy [mm]	±10 (@ 100 % target reflectivity)
Frame rate [fps]	Up to 54 (depending on the camera settings)

carried out for architectural survey purposes in order to understand the possible problems that could arise concerning a decrease in accuracy during a survey on real objects.

2.1 Pre-calibration and Data Acquisition Methodology

According to some previous works reported in [9, 15, 39, 41], TOF camera measurements are affected by the internal temperature of the measurement system; for this reason, a warm up period of the camera of least 40 min is necessary.

After the warm up, the camera, mounted on a topographic or photographic tripod, acquires at least thirty frames with an integration time that is equal to the "auto integration time" suggested by the SR_3D_View software. These frames are than averaged in order to reduce the measurement noise.

After averaging the acquired frames for each camera position, each pixel distance measurement is corrected using a distance error model [9] in order to obtain the final range images that are useful for the survey purposes. This procedure has been used in all the tests and applications that are presented in the following sections.

Before starting with applications on real objects it is necessary to evaluate the performance of the TOF camera, considering the influence of the incidence angle and the influence of object reflectivity on distance measurement accuracies: both phenomena represent typical conditions that occur when the objective of the survey is an architectural object or an artefact (usually built with different materials and surfaces).

2.2 Influence of the Angle of Incidence on Distance Measurements

The signal emitted by the camera impinges the observed object with an angle which depends on the camera orientation with respect to the normal of the object surface. We can define α as the angle between the optical axis of the camera and the normal to the object surface, as shown in Fig. 1.

Some works have already examined the influence of the emitted signal angle of incidence on distance measurement precision [2, 18]. In the following, this aspect is analyzed from a more practical point of view: the analysis deals with data acquired with the SR_3D_View software using the "auto integration time", thus changing the integration time for each object position, as a generic user could do, and acquiring data of the object to be surveyed.

In order to evaluate whether there is any influence of the α values on the precision of the distance measurements acquired in this way, the following system is considered.

The camera was positioned on a photographic tripod, with the camera front parallel to a Plexiglas panel covered with a white sheet, which was fixed to the top of a total station (Fig. 2, left). After the camera warm up, the panel was accurately rotated using the total station, by two degrees at a time in the $0° \div 45°$ range (Fig. 2, right), in both clockwise and anticlockwise directions, while the SR-4000 camera was fixed. Fifty consecutive frames were acquired for each panel position, using an integration time equal to the "auto integration time" suggested by the SR_3D_View software. The distance between the panel and the camera was about 1.6 m.

In order to accurately estimate the distribution of the distance measurements around their mean value, a reference plane was estimated for each panel position, after outlier elimination, from the acquired range images using a robust estimator (Least Median Squares (LMS), [37]).

This estimator has a high breakdown point, which means that it can discriminate outliers and leverage points which can be up to 50 % of the considered data.

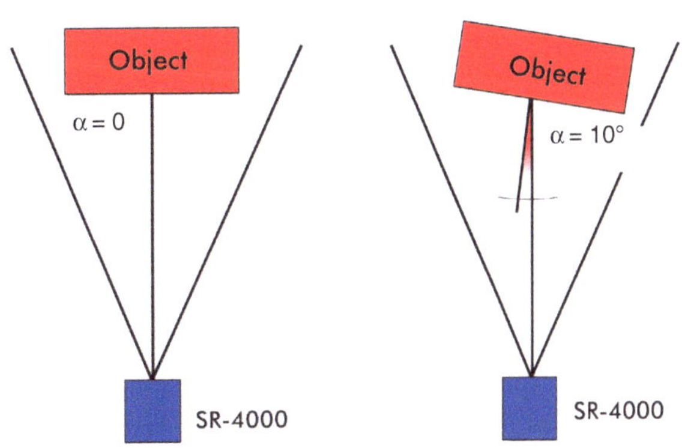

Fig. 1 Alpha angle between the optical axis of the camera and the normal to the object surface

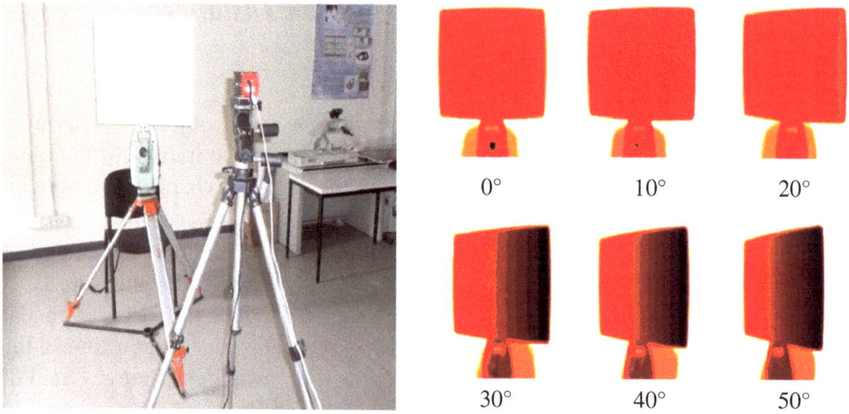

Fig. 2 System used to evaluate the influence of the alpha angle on camera distance measurements (*left*); some positions of the panel during the acquisition (in *false colour right*)

The parameter which has most influence on the LMS is the threshold value of rejection L, which offers a preliminary hypothesis on the percentage of outlier contamination. After testing this estimator on several randomly generated range images, containing different percentages of outliers, we adopted a threshold value of rejection of $L = 1.5$.

The LMS estimator was applied to a sub-image of 65×61 pixel dimensions, which was centred with respect to the centre of the panel in each position. Using this estimator, it was possible to select some reliable points in the sub-image that were necessary for a robust plane estimation. The differences between the range image (obtained after averaging fifty frames) and the estimated reference plane were then calculated for each panel position, always considering the sub-image of 65×61 pixel dimensions. The mean and standard deviation values of these differences are reported in Fig. 3 on the left and right, respectively. In the case of $\alpha > 45°$, the panel area was too small for a reliable estimation of a reference plane, therefore the analysis was limited to 45° in both directions.

The left side of Fig. 3 shows that the mean value of the differences between the estimated plane and the SR-4000 distance measurements undergoes small fluctuations around the zero value according to the α values: these small fluctuations are less than 2 mm in both the clockwise and anticlockwise directions. The standard deviation values vary according to the α values (Fig. 3 right): this variation is limited to about 2 mm. This trend is justified by the adopted procedure: since the data were acquired with the "auto integration time" mode for each panel position, the reduction in the amount of reflected light from the panel is limited to about 20 %, with respect to the reflected light from the initial position ($\alpha = 0°$). The standard deviation of distance measurement is in inverse proportion to the amplitude of the reflected light [4, 7, 39]; therefore, an amplitude reduction of about 20 % will result in an approximate increment of about 25 % of the standard deviation of the distance measurement. Since the typical standard deviation value

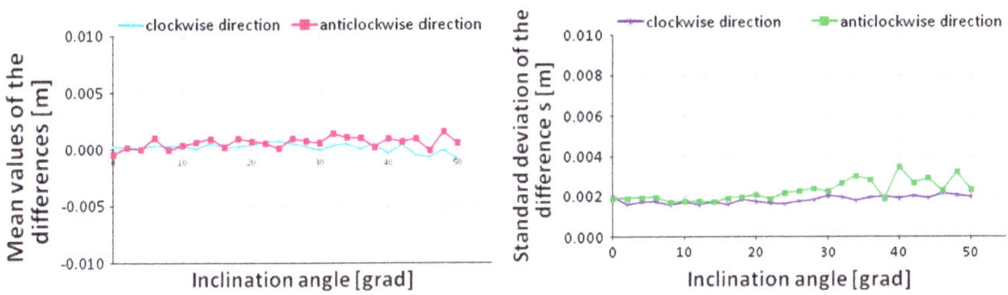

Fig. 3 Mean values of the differences between the range image and the estimated reference plane (*left*); standard deviation values of the differences between the range image and the estimated reference plane (*right*)

of the distance measurements is 4 mm (see www.mesaimaging.com), a 25% increment of that value can be considered negligible. This aspect can be confirmed from the previously reported results. In conclusion, if the "auto integration time" is adopted for data acquisition, there is no appreciable variation in the distance measurement accuracy for camera orientations that fall within the considered interval of α values.

2.3 Influence of Object Reflectivity on Distance Measurements

The standard deviation of distance measurement is in inverse proportion to the amplitude of the reflected light, which in turn depends on the reflectivity of the object to the signal emitted by the camera when all the other parameters (integration time, distance between camera and object, background illumination) are kept constant. Therefore, a study on the influence of object reflectivity on distance measurements is necessary in order to investigate the potentiality of the camera for architectural metric surveys. For this purpose, the system represented in Fig. 4 left was considered.

This system allows planar objects of different dimensions to be positioned in order to be tested, and offers high stability compared to displacements induced by changing the objects, thanks to the use of appropriate supports. The camera is positioned on a photographic tripod, parallel to the wooden panel of the system.

The materials to be tested are positioned from one at a time on the system, while the camera is fixed. Fifty frames were acquired for each material (Fig. 4, right) and then averaged in order to reduce the measurement noises. This procedure was repeated with several distances (from 1.30 to 1.80 m) between the camera and the tested objects, moving the camera rather than the system.

The camera positions (defined using 5 points) and the object surface positions (defined using 6 points—see Fig. 5, left) were estimated in an arbitrary coordinate

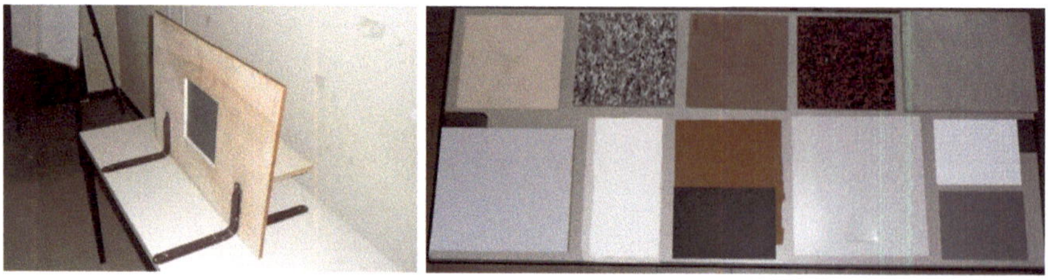

Fig. 4 Purpose-built system used to position planar objects of different dimensions (*left*); some of the tested material (*right*)

system thanks to accurate topographic measurements, using two total stations (Fig. 5 right) in a direct intersection scheme.

The tested materials and the integration times (I.T.) adopted for the data acquisition in the considered case (1.80 m from the camera) are reported in Table 2: the tested planar objects were chosen from among common building materials which can be found in the case of both indoor scene reconstruction and architectural element surveying. The same tests can be run using the specific materials of the object that have be surveyed.

Only the data acquisition and processing details relative to a distance of 1.80 m between the camera and system are reported in Table 2. Fifty frames were acquired twice for each material, with two different integration times: "I.T. auto" ("auto integration time"), which shows small variations that depend on the material reflectivity and "I.T. ref.", which corresponds to the "auto integration time" for the Kodak R27 grey card, and which was adopted as the reference integration time for the considered distance. In this way, it was possible to compare the reflectivity of the tested materials with the "standard reflectivity" obtained for the Kodak R27 grey card.

In order to avoid noise effects, caused by the presence of the wooden panel and of the depth discontinuity between the wooden panel and the surface to be tested,

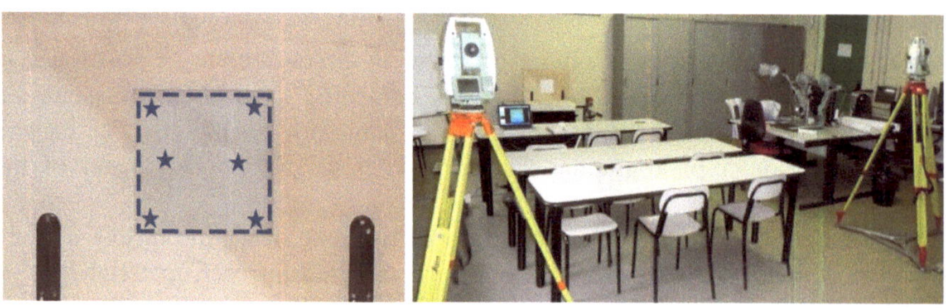

Fig. 5 Points measured for each material in order to estimate the spatial position (*left*); Topographic measurements for the estimation of the camera and tested object positions (*right*)

Table 2 Results considering data acquired in "I.T. auto" mode (1.80 m)

Material [-]	I. T. auto [ms]	I.T. ref. [ms]	Mean of the differences [m]	St.d. of the differences [m]
Kodak R27 dark	106	106	0.002	0.001
Kodak R27 bright	102	106	0.001	0.001
Hardboard	107	106	0.002	0.001
Black paper	107	106	0.001	0.001
Laminated wood	105	106	0.006	0.001
Bright plasterboard	104	106	0.001	0.001
Painted metal sheet	99	106	0.008	0.001
Marble Pietra Etrusca	108	106	0.002	0.001
Balmoral red granite	107	106	0.003	0.721
Granite	107	106	0.004	0.781
Marble Pietra Orsera	105	106	0.007	0.739
Stone	107	106	0.000	0.001

the analysis of the SR-4000 distance measurements was limited to a square area inside the surface of the tested materials. The differences between the estimated plane (using the points acquired with the topographic survey) and the camera distance measurements were estimated for each of these materials, considering both "I.T. auto" and the "I.T. ref.". The mean and standard deviation values of the differences for "I.T. auto" are reported in Table 2. Similar results were obtained for "I.T. ref.".

It is possible to observe, in Table 2, that the mean value of the calculated differences shows small variations, between the tested materials, which are of the same order as the camera distance measurement accuracy. It is worth nothing that the mean value of the differences (raw data accuracy) of all the considered materials is 3 mm; the same variations are observed using "I.T. ref.". The standard deviation value of the calculated differences is less than 2 mm for all the materials, except for "Balmoral Red Granite", "Antigorio Scuro" and "Marble Pietra Orsera". For these materials, in fact, there was a high percentage of saturated pixels, due to their high reflectivity to the camera signal, and the distance measurements were quite heterogeneous, but after the elimination of the saturated pixel, the values of the Standard Deviation of the differences concerning the three previously mentioned materials returned to very similar values to the other ones (0.51, 0.53, 0.14 cm respectively).

On the basis of the obtained results, it is possible to state that there is no decrease in the camera accuracy of distance measurements for the typical building materials employed in the architectural Cultural Heritage.

2.4 Tests, Data Processing, Comparisons and Results in Cultural Heritage Architectural Surveying

Some results are here presented in order to show the potentiality of TOF cameras in Cultural Heritage architectural surveying and of their integration with other techniques. First, a complete survey of an architectural frieze was conducted (Fig. 6, left) and the TOF camera accuracy was evaluated with a Laser Scanner survey on the same object (60 × 20 × 40 cm). Moreover, two windows (3 × 6 m) containing different materials (Fig. 6, centre and right), were surveyed in order to create 2D architectural drawings and 3D models which are useful to obtain knowledge and the documentation of the objects. The results, the encountered problems and the final results are reported hereafter.

2.5 An Architectural Frieze, Survey and Accuracy Evaluation of TOF Camera Performances

As can be seen in Fig. 7 (left), the frieze was positioned on a table in front of the SR-4000 camera at a distance of 2 m (Fig. 7, centre).

Seven purpose-built cubic targets, covered with a white sheet, were distributed around the object that had to be acquired in order to obtain some reference points which are necessary to define a known coordinate system for the two acquired datasets.

The camera was mounted on a photographic tripod in front of the frieze and fifty frames were acquired with the TOF camera adopting the auto integration time suggested by the SR-3D-View software and then averaged in order to reduce measurement noise. The Z coordinate (orthogonal distance between the front of the camera and the object) of the surveyed points was then corrected with the distance calibration model [9] in order to obtain more reliable coordinates.

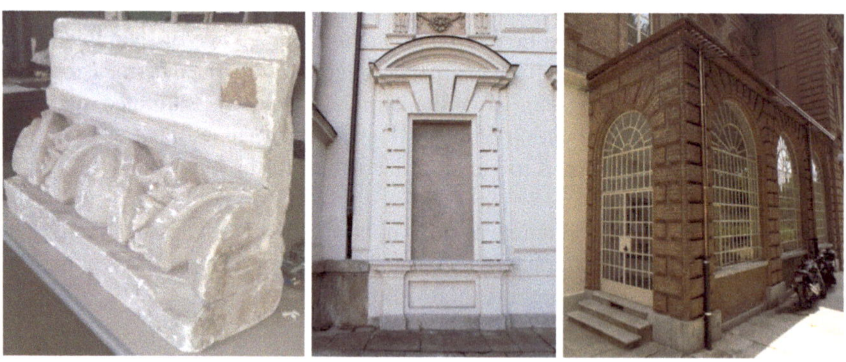

Fig. 6 The architectural frieze (*left*); windows of the Valentino Castle (*centre* and *right*)

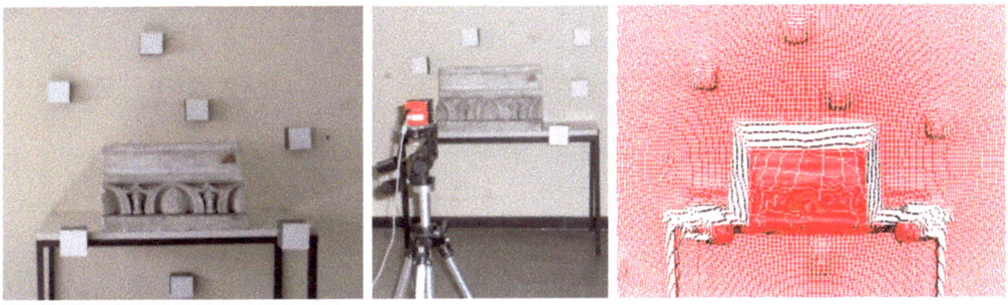

Fig. 7 The position of the architectural frieze (*left*), data acquisition with the SR-4000 camera (*centre*), and the acquired TOF point cloud (*right*)

The limited dimensions of the architectural frieze made it possible to acquire the entire shape of the object with a single acquisition (Fig. 7, right). The step between the points using the SR-4000 camera at 2.0 m was about 3 mm, which can be considered a good rate to obtain a complete 3D model of the object. The different products that were achieved are reported in the following figures (Fig. 8).

A complete survey was conducted after the TOF data acquisition using the MENSI S10 triangulation based laser scanner in order to test the accuracy of the camera for real object survey (Fig. 9, left). As the S10 scanner has sub-millimetre accuracy, the data acquired with this instrument (Fig. 9, centre and right) can be used as a reference since the MENSI S10 accuracy is less than 1/10th of the expected accuracy of the SR-4000 camera.

In order to perform the comparison between the two "point clouds", the data acquired with S10 were inserted into the same coordinate system as the TOF camera using the previously mentioned cubic targets.

Finally, the difference between the distance of the corresponding point in the MENSI S10 data (reference data) and the distance measured by each pixel of the SR-4000 camera (TOF original data) were estimated.

Figure 10 shows that the estimated differences vary when objects that are at different distances from the camera are considered since the error function depends on the distance between the camera and the object. Since the SR-4000 data and the MENSI S10 data were acquired from slightly different positions, some areas in Fig. 10 show very different values, which can be considered wrong. The mean value of the differences in the object area considering the original TOF data was

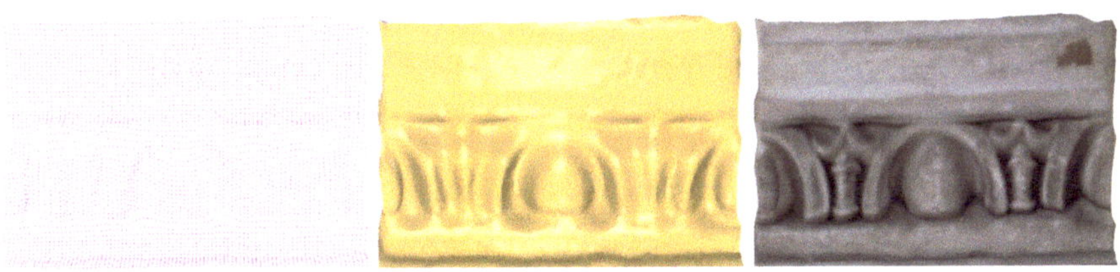

Fig. 8 Point cloud (*left*); mesh (*centre*); 3D textured model of the surveyed architectural frieze (*right*)

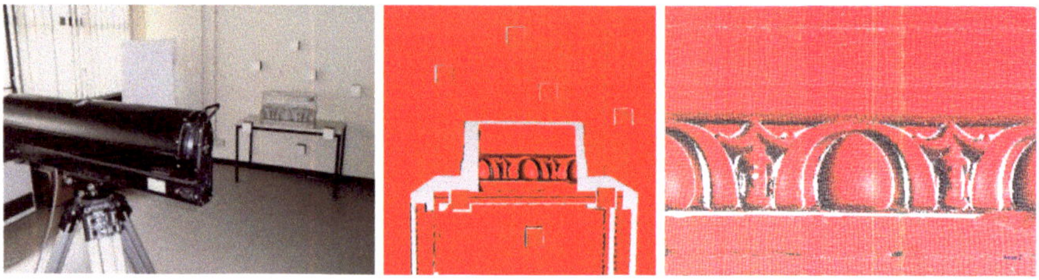

Fig. 9 Data acquisition of a frieze with the MENSI S10 laser scanner (*left*); the complete acquired point cloud (*centre*) and details of the decoration (*right*)

Fig. 10 Differences (expressed in metres) between the distances obtained from the Mensi S10 point cloud and the SR-4000 point cloud

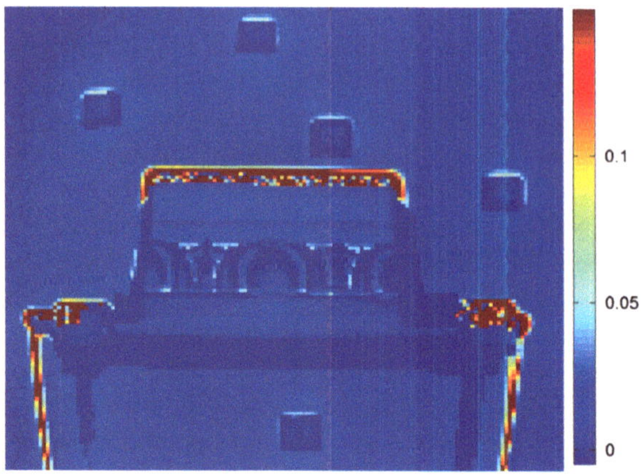

0.6 cm, instead when a distance calibration model is applied [9], the mean value of the differences became 0.1 cm. These results demonstrate the efficacy of the distance calibration model and show the high potentiality of using TOF cameras for metric surveys of architectural artefacts without strong break-lines and for 3D object reconstruction.

2.6 Data Acquisition and Processing of Two Windows

After the test on a real object in indoor conditions a first outdoor test was carried out concerning a survey on two windows pertaining to the Valentino Castle, Turin, Italy (one in the main building and the other in the Chevalley building). The aim of this test was first to evaluate the performance of the TOF cameras during a complete architectural survey of an object which required strong break-lines to describe it correctly and, at the same time, to estimate the capability of these instruments in outdoor conditions.

After the camera warm-up, the SR-4000 camera was positioned on a photographic tripod and moved to different positions in order to achieve complete coverage of both windows (Fig. 11).

Some problems emerged in the acquisition data (pixels too saturated) during the central hours of the day, when the sunlight illuminated the objects of the survey. For this reason, the surveys were performed in the early hours of the day and in the late afternoon, after the sun had gone down. This problems has shown that at the moment it is not possible to acquire reliable range images with the camera that was employed when constant, persistent and direct sunlight is present.

When the environment conditions were favourable, thirty frames were acquired from each position using the software supplied with the camera (SR_3D_View software), and adjusting the integration time in order to obtain the minimum possible number of saturated pixels and a low noise level of the distance measurements (when the "auto-integration time" suggested by the software was used too many pixels were saturated and, as a consequence, the data were not suitable for use).

The TOF data were acquired from several different positions in order to obtain a complete 3D model of each window and the acquired range images were overlapped by about 50 %.

Six range images were acquired for the first window (Fig. 11, left) at a taking distance of between 2 and 4 m. The step between each point of the recorded point clouds was about 15 mm with acquired area dimensions of about 3.00 × 2.50 m for each range image.

The thirty frames acquired from the different positions were then averaged as is usual practice. The distance recorded for each pixel of the averaged frames was then corrected with the same distance error model used in the previous tests. Some range images of the first acquired window are reported in Fig. 12.

The second window (Fig. 11, right), which is characterized by a brick decoration and has a very large glass surface, was acquired using the same previously described strategy. In this case, eight range images were acquired (Fig. 13).

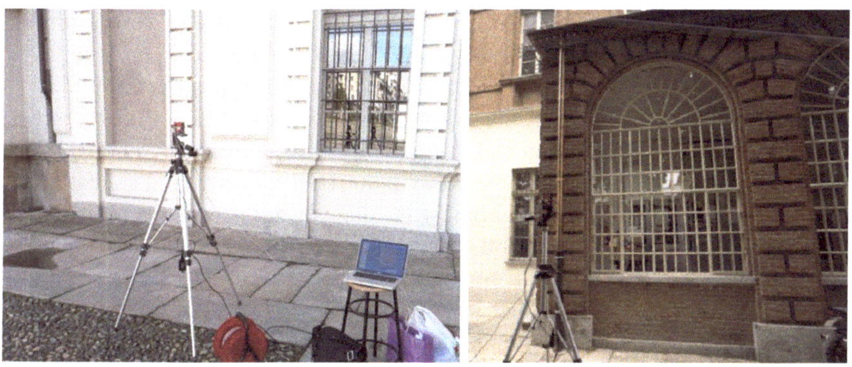

Fig. 11 TOF data acquisition of the two windows at the Valentino Castle

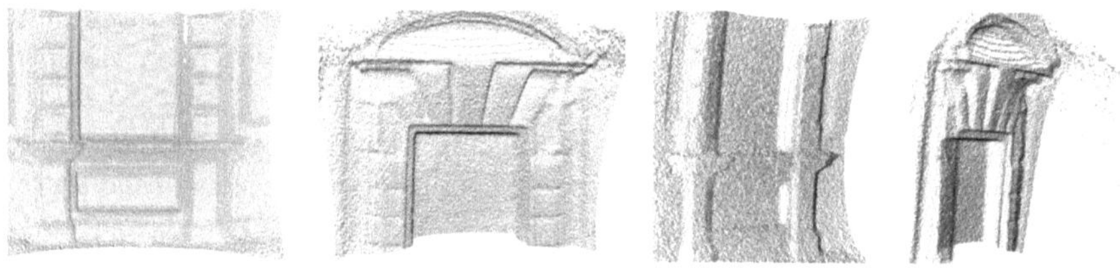

Fig. 12 Three parts of the first window surveyed using the TOF camera

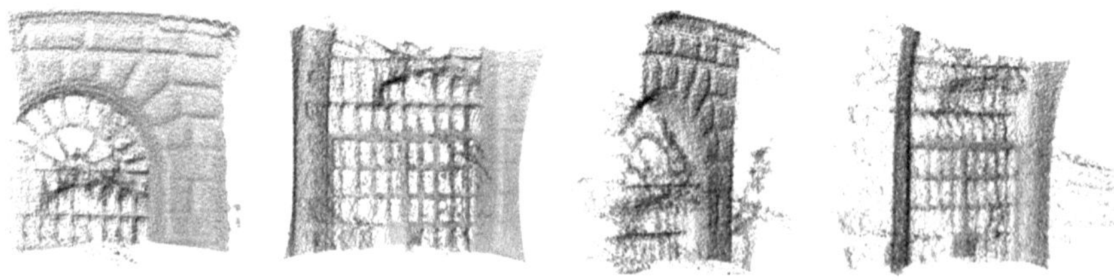

Fig. 13 Three parts of the second window surveyed using the TOF camera

On the basis of the resolution of the camera, the step between each point was about 15 mm.

After the first processing steps, which allowed the TOF scans to be obtained without any systematic errors, the next phases were focused on obtaining the complete model of the windows. Adopting the typical approach used for Laser Scanner data processing, all the point clouds were first cleaned (the outliers and the occlusions were deleted), and then the range images were registered using the ICP algorithm (Geomagic Studio Software was used) in order to obtain a 3D model for each window. This process is time consuming because, during the first step, it is necessary to manually measure the homologous points in each scan, and after this first approximation, the algorithm is then able to start and to conduct the registration. The final models of the surveyed windows are reported in Fig. 14.

The registration process of the first window was successful (average discrepancies on check-points were of about 4 mm) thanks to the excellent definition of the geometric shape of the surveyed windows (Fig. 14, left and centre).

However the ICP algorithm did not delivery very good results for the second window; the shape of the windows (only a few geometric edges) and the large windows panes, which increased the noise of the TOF data, required a great deal of manual effort (more than a few homologous points were necessary) to realize the final point cloud. Despite the problems encountered regarding the second window, the final result was acceptable (average discrepancies on check-point were 25 mm) and a 3D modelling was obtained.

The final point clouds were imported into a modelling software and some tests were conducted on the modelling of the object in order to analyse the suitability of

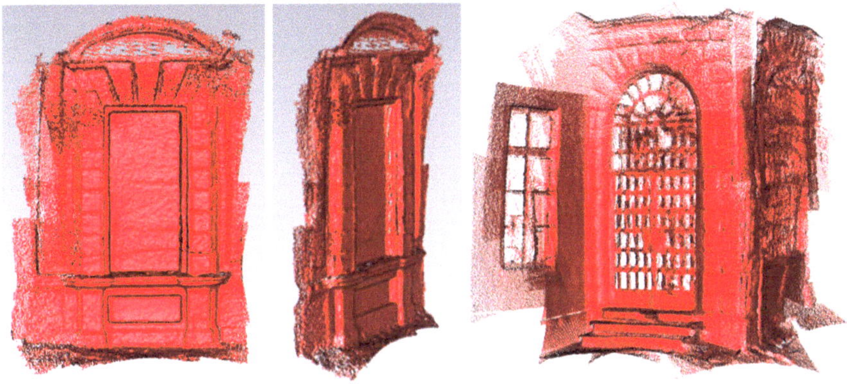

Fig. 14 Screen-shots of the complete 3D TOF point cloud (first window *left* and *centre*, second window *right*)

these data for a correct 3D model (Fig. 15) and for break-line extraction to obtain typical 2D representations.

As can clearly be seen from Fig. 15, on the left, it is not possible to recognise the real shape of the surveyed object from the model; in this case, the problems arose because of the geometry of the window. The shape of the plaster (the main decorative feature of the window) in particular shows several well defined edges and when the aforementioned point cloud step (15 mm) was introduced, the geometry was not recognizable from the TOF camera data. As a result, the 3D model was too smoothed and, when only the data obtained from the camera were used, it was impossible to reconstruct the correct shape of the windows.

The model of the second window (Fig. 15, right) also failed to represent the correct shape of the surveyed object. In this case, the problem was the aforementioned high noise of the TOF point cloud, which produced a noisy 3D model. In this case, during the modelling phase, some attempts could be made to reduce the noise rate, but this would result in a large smoothing of the window which would loose all of its characteristic shape. The model was not able to show the real profile of the surveyed object.

On the basis of the obtained results, it is possible to state that in the case of objects with well-defined break-lines, it is not possible to obtain the typical products of an architectural survey, such as 2D drawings or 3D models using only the data acquired with the TOF camera.

For these reasons, the approach followed in the second part of the test was focused on the integration of the TOF data with digital photogrammetric data derived from multi-image matching (break-line extraction), in order to obtain the required products and to obtain a more accurate shape description of the object that would be useful for the creation of architectural products (drawings and 3D models).

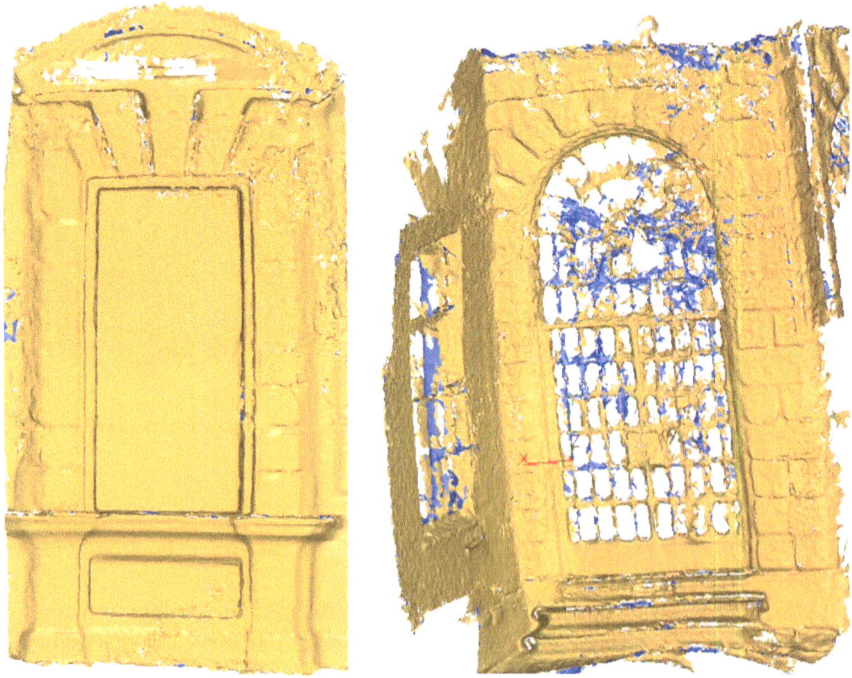

Fig. 15 The 3D model obtained using the TOF data of the window of the main building of the Valentino Castle (*left*) and the Chevalley building window (*right*)

2.7 Digital Photogrammetry Multi-Image Matching Approach for Break-Line Extraction

It is well known that in order to improve the potentiality of an image-matching algorithm for break-line extraction it is important to have an approximate surface available of the object that has to be surveyed. The proposed approach has therefore been based on the use of TOF data, during the multi-image matching data processing, in order to provide an accurate digital surface model of the aforementioned extraction.

The algorithm which can be summarized in several steps, is shown in Fig. 16 left (the additional information of the TOF camera for the multi-image matching approach is highlighted in dark grey).

The images were acquired adopting an ad hoc taking configuration (Fig. 16, right): several images were acquired and the most central one was considered as a reference image during the matching process. The TOF point cloud was registered from a central position, with respect to the image acquisition, in order to have approximately the same occluded areas in the TOF data and in the reference image.

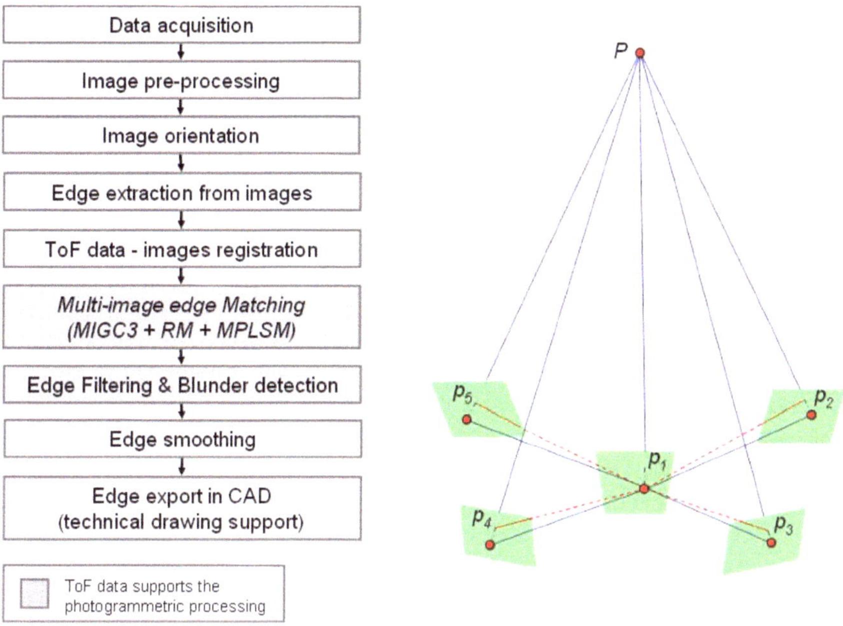

Fig. 16 Workflow of the break-line extraction (*left*) and ad hoc image taking configuration (*right*)

All the images were previously restored and then enhanced. The image restoration was performed by means of Adaptive Gaussian Smoothing [30] which filters the image according to the noise level evaluated on the uniform areas of the image. Image enhancement was then obtained using a Wallis filter [40]. Image pre-processing generally allows the number of detected edges to be increased, compared to the original image.

The orientation was performed in a proper reference system in order to have the z-coordinate normal to the main plane of the façade. In this step, the A^2SIFT (Auto-Adaptive Scale Invariant Feature Transform) operator [27] was adopted in the tie-point extraction, and a robust (Least Median Square) relative orientation was then performed in order to eliminate the mismatches [26]. Finally, a bundle block adjustment was performed. The edge extraction was then performed using the Canny operator [8] on the reference image. The extracted edges were then approximated, by identifying the pixels in which the edge changed in direction as knots and linking these dominant points with straight lines.

The point cloud was registered in the photogrammetric reference system using a spatial similarity transformation. In this way, it was possible to share the information between the images and the point cloud. A multi-image matching algorithm was then set up. The algorithm is a modification of the Geometrically Constrained Cross Correlation (GC^3) [42]: it uses a multi-image approach, that is, it considers a reference image and projects the image patch (of each dominant point) of the reference image onto the DSM (TOF point cloud), and then, using the approximate z-value achieved by the DSM, back-projects it onto the other images. Through this algorithm, the dominants points of each edge were matched in all the images in order to reconstruct the break-line positions in 3D. The images were preliminarily undistorted (using the camera calibration) in order to ease them into a central perspective. The epipolar constraint limited the search space in the images. The length of this line was achieved considering the z-value given by the TOF point cloud; then, in order to find the homologous point in all the images, this value was varied in a Δz range. This work was enforced and improved through the use of the position of already matched points: the z-value of two adjacent dominant points on the same edge had to be similar. In this way, it was possible to reduce the run of the epipolar line on the façade to a few centimetres. A relational matching was developed to improve the rate of the successfully matched points. This algorithm integrates the figural continuity constraint through a probability relaxation approach [10] and is able to solve several ambiguities of the matching process. The method uses the already matched dominants points as anchors and, in an iterative way, defines the most suitable match between candidates imposing a smoothing constraint. Finally, a Multi-Image Least Square Matching (MILSM) [3] was performed for each extracted point, in order to improve the accuracy to a sub-pixel dimension.

Some blunders were generated during the matching process. These blunders were first deleted from the extracted edges using a filter which considered the reciprocal point positions on the same edge: the position of a point was predicted considering the neighbouring points of the edge and the difference between the predicted and the real position of the point was then evaluated. If the difference value was higher than a predefined threshold, the point was deleted. This filter is not robust: it works well if the blunders are isolated from each other. For this reason, a second filter could be used to clean the edges when several blunders are close together; this algorithm uses the TOF information to verify the correctness of each dominant point: when it is further than a defined threshold from the point cloud, it is deleted [30].

Image matching allows radiometric edges to be extracted. Most of these edges are due to shadows or radiometric changes, but they do not have a geometric correspondence. Only geometric boundaries are of interest in the surveying of graphic drawings and for modelling purposes. For this reason, the position of each dominant point on the extracted edges was considered with respect to the TOF point cloud: it was verified whether a geometric discontinuity occurred in the TOF data close to the edge point.

The edges extracted by the matching algorithm were affected by random noise and they could not be used directly in the drawing production. For this reason, the noisy edges were split into basic elements (linear and curved elements) and each element was smoothed and eased, in an automatic way, into lines and second-order curves by means of a polynomial fitting. The basic elements were then recollected in a single smoothed edge [31].

Finally, the geometric edges were exported into CAD in order to obtain preliminary data for the graphic drawings of the survey and for a rough evaluation of the achieved results.

2.8 Results and Discussions

Starting from the images acquired using a calibrated Canon Eos 5D digital camera with a 24 mm lens, in the ad hoc configuration reported in Fig. 16 right, the images of the two windows of the Valentino Castle (Fig. 17) were processed using the proposed image matching algorithm.

The integration between the TOF data and the digital images was performed according to the aforementioned workflow (Fig. 16, left). The edge extraction allowed a complete set of lines to be defined from the reference image of each case: Figs. 18 and 19 shows the extracted edges, which could be useful for the realization of to obtain the final architectural products.

In order to achieve a complete product of the surveyed object, the last phase of the data processing is usually the production of 2D drawing and/or 3D models to improve architectural knowledge and documentation. These products, which allow a semi-realistic view of the object, can be derived in different ways, such as using the principal geometry of the object [21, 28], derived from a photogrammetric plotting, or a traditional topographic survey, or even from LiDAR data [5, 6, 35] or using other instruments that allow 3D point clouds to be obtained (for instance TOF cameras).

In the present case, the approach was based on the integration of the smoothed edges (break-lines), obtained after multi-image matching, with the 3D data derived from the TOF camera.

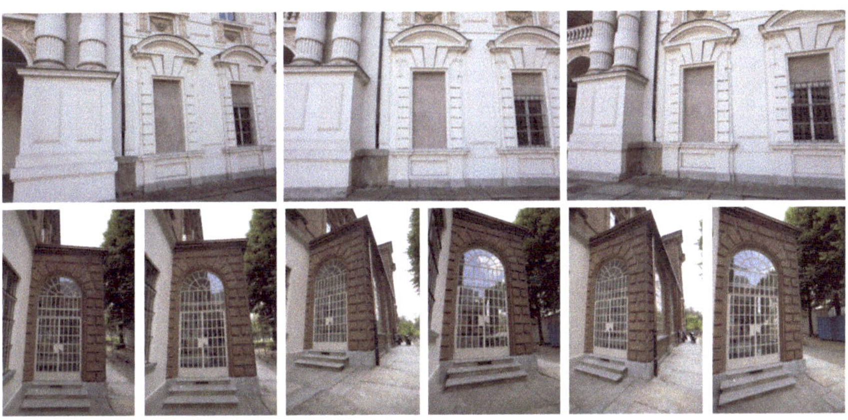

Fig. 17 Some images acquired for the multi-image matching process (*Above*: three shots of the first window; *Below*: six shoots of the second window)

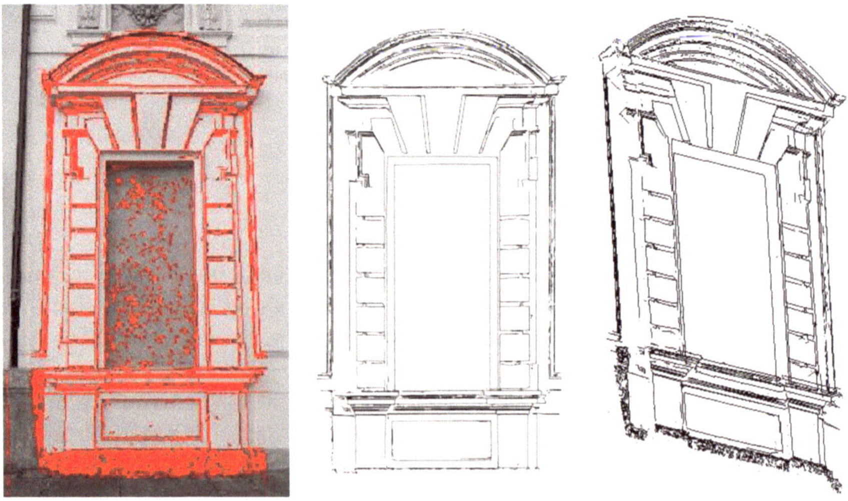

Fig. 18 The window of the main building of the Valentino Castle: extracted edges on the reference image (*left*) and smoothed edges (*centre* and *right*)

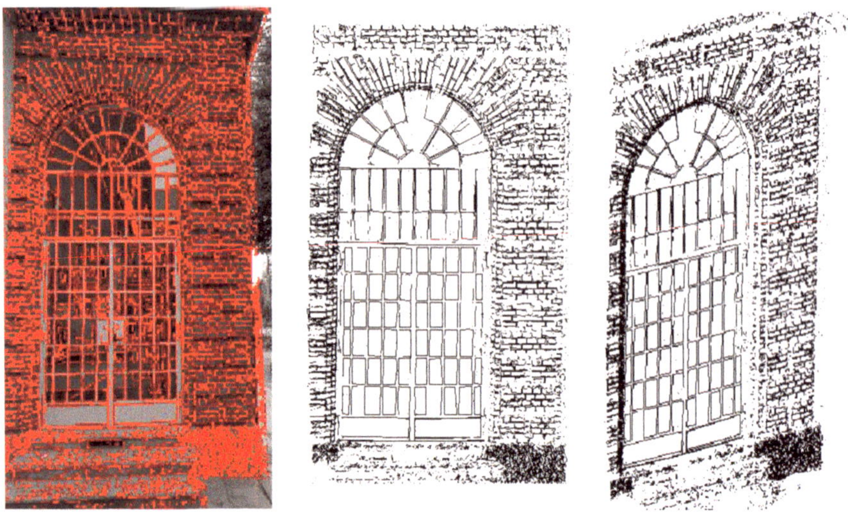

Fig. 19 The Chevalley building window at the Valentino Castle: extracted edges on the reference image (*left*) and smoothed edges (*centre* and *right*)

The obtained data (break-lines and TOF point cloud) were first integrated (Fig. 20 left), and the modelling phase was then realized. In this way a complete 3D model was obtained (Fig. 20 centre and left). Finally, the model was textured in order to create some photorealistic views of the object (Fig. 21).

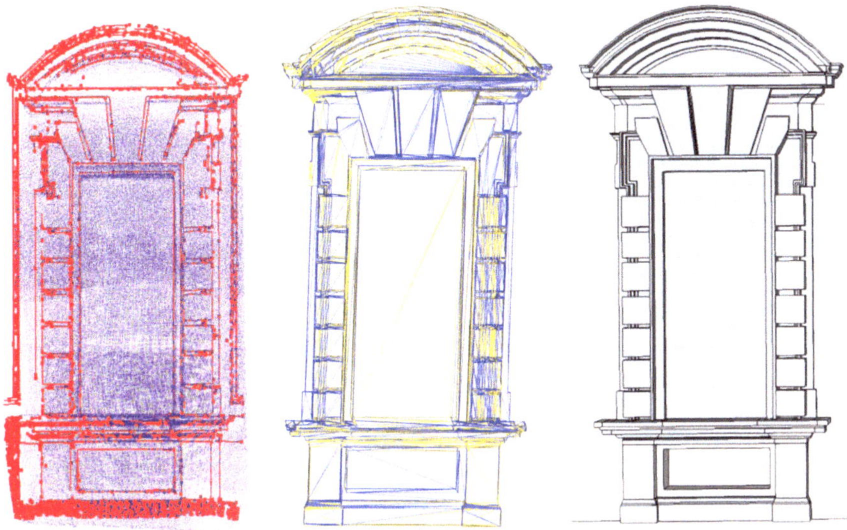

Fig. 20 TOF and break-line data integration (*left*); views of the final 3D model (wireframe visualization, *centre*, triangulated surface, *left*)

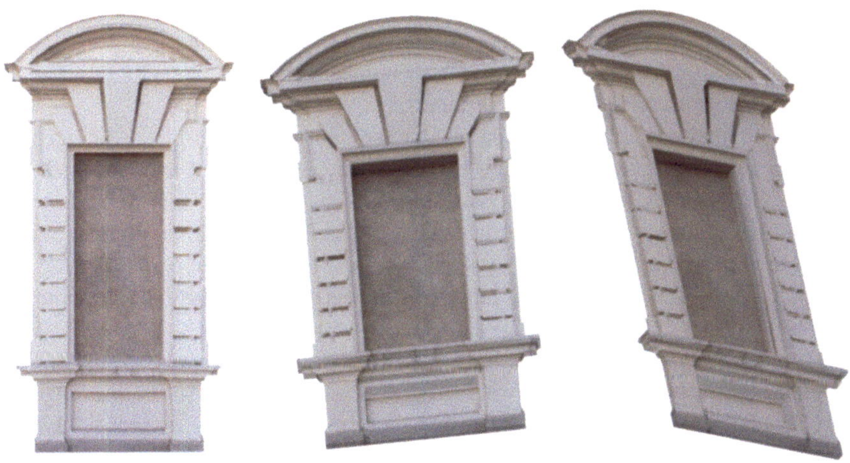

Fig. 21 Photo-realistic views of one of the surveyed windows

Another fundamental product that is required as a final representation, after an architectural survey, is a typical 2D representation. Whatever techniques is used for the survey: 3D data acquisition systems, photogrammetric image matching techniques or typical topographic systems (Total stations etc.), the most important element that allows correct and comprehensive drawings to be obtained, is the

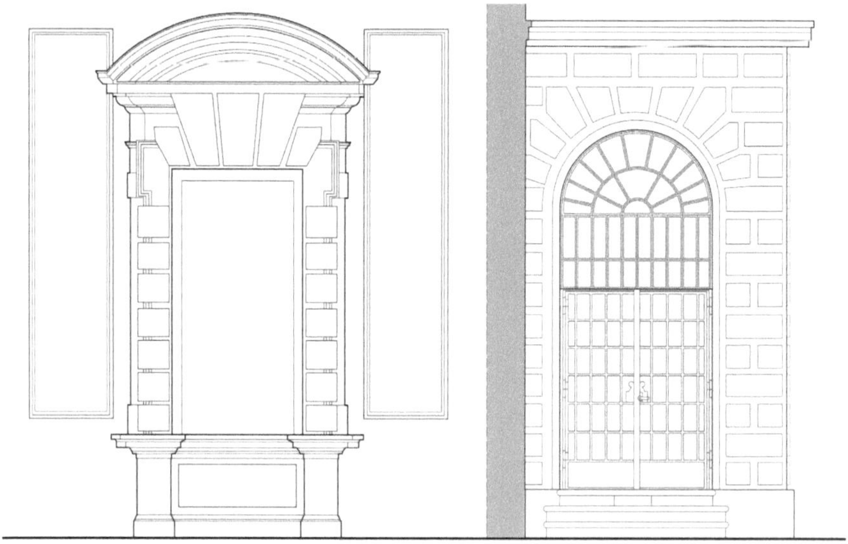

Fig. 22 The final 2D representation of the surveyed windows

knowledge of the architecture. It is therefore important to involve several professional figures (Historians, Architects, Engineers) in an architectural survey in order to achieve correct metrical products.

It is of fundamental importance to analyse and study the object. An accurate knowledge of the geometric shape of the object is necessary and, using all the acquired cognitions, it is finally possible to realize a final 2D representation.

The methodology reported in the examples was based on this multidisciplinary approach and, thanks to the involved various forms of expertise, using the extracted break-lines, the TOF data and the derived models the following 2D representation were realized (Fig. 22).

The accuracy of the final drawings and 3D models was checked using several ground control points acquired with a Total Station. The measured discrepancies (less than 2 cm for each point) show that these representations respect the typical precision of a drawing at a 1:100 scale.

The achieved products shown that, through an interesting integrations of different survey methodologies it is possible to obtain complete metric documentation of an architectural object. However it is important to underline that, at the moment, the low resolution of the employed TOF camera (SR-4000) does not allow autonomous surveys to be performed at a typical architectural scale (1:100–1:50) and that it is necessary to improve the achieved TOF data of these devices with other metrical information derived from different techniques or instruments.

3 Conclusions

Architectural metric surveying essentially requires the creation of 2D representations in order to describe an object, but in recent years an important increase has been witnessed in the realization of 3D models for architectural knowledge and documentation purposes.

A geometric break-line definition is needed to create traditional 2D drawings while point clouds of regular surfaces are generally not useful: point clouds can only satisfy metric survey requirements when irregular and smoothed surfaces have to be described. Only high resolution point clouds, acquired using last generation TLS or multi-image matching approaches in digital photogrammetry, seem be able to give better answers at medium scales (e.g. 1:100).

Unfortunately, as it is well known, although these new instruments and methodologies can speed up the acquisition phase and register a great deal of information very quickly, the processing of point clouds for the creation of 2D representations is not easy and is very time consuming.

In the past, the selection of points by a human operator using total stations, distance measurements or photogrammetric plotting, made it necessary for the user to select only the necessary information during the acquisition and processing phases: the geometric points and break-lines that delimitated the surveyed objects were usually selected.

TOF data are usually affected by some systematic errors that can be partially corrected using a suitable calibration procedure. In this way, complete 3D point clouds are obtained that are comparable with those of traditional TLS acquisitions.

When the object has smoothed and irregular surfaces (see previously analysed architectural frieze), the results are excellent and could be considered suitable to obtain knowledge of small objects and for documentation purposes (e.g. decorative features, archaeological relicts, etc.). Instead, when the dimension of the object increases and the geometry is well defined (which is typical in Cultural Heritage architecture), and a better definition of the break-lines is necessary, TOF data alone are not enough to set up 3D models or 2D drawings at an architectural scale.

For these reasons, the integration of these data with a multi-image matching approach is necessary. Using this approach, it is possible to drastically reduce both the data acquisition and processing times for 2D and/or rough 3D drawing generation. Moreover, a combination of the two techniques allows typical and metrically correct architectural representation to be obtained.

On the basis of the encouraging results that have been obtained, the necessity has emerged, of having to refine some procedures and processing steps; first the acquisition phase and the registration process need to be improved (for instance, using the proposed approach for automatic registration and mixed pixel removal of TOF data [32]). The performances of the integration algorithm also need to be refined to improve the completeness of the achievable results obtained from the multi-image matching techniques.

Considering the possible and foreseeable upgrading of the TOF camera hardware, more resolution and larger taking distances are necessary, and the possibility of recording RGB information on the same CCD array (or in a different one with a calibrated location inside the same body) is one result metric surveyors are waiting for.

Currently, the quality of a single frame does not seem able to a give correct response, therefore TOF cameras are used in a static way. The development of ad-hoc software and hardware solutions to reduce and eliminate noises and the registration strategies of each frame with the adjacent ones will allow the TOF camera to be used as a 3D video camera, which could speed up point cloud acquisition in a significant way for all indoor applications.

References

1. M.A. Albota, R.M. Heinrichs, D.G. Kocher, D.G. Fouche, B.E. Player, M.E. Obrien, G.F. Aull, J.J. Zayhowski, J. Mooney, B.C. Willard, R.R. Carlson, Three-dimensional imaging laser radar with a photon-counting avalanche photodiode array and microchip laser. Appl. Opt. **41**, 7671–7678 (2002)
2. D. Anderson, H. Herman, A. Kelly, Experimental Characterization of commercial flash ladar devices. In: Proceedings of International Conference on Sensing Technologies, Palmerston North, New Zealand, (2005)
3. E. Baltsavias, Multiphoto geometrically constrained matching. Ph.D. Dissertation, ETH Zurich, Switzerland, ISBN 3-906513-01-7 (1991)
4. N. Blanc, T. Oggier, G. Gruener, J. Weingarten, A. Codourey, P. Seitz, Miniaturized smart cameras for 3D-imaging in real-time, in *Proceedings of IEEE Sensors*, (Vienna, Austria, 2004), pp. 471–474
5. J. Böhm, Terrestrial laser scanning—a supplementary approach for 3D documentation and animation, in *Photogrammetric Week '05*, ed. by Fritsch. (Wichmann, Heidelberg, 2005), pp. 263–271
6. C. Bonfanti, F. Chiabrando, A. Spanò, High accuracy images and range based acquiring for artistic handworks 3D models. Int. Arch. Photogram. Remote Sens. Spat. Inf. Sci. **XXXVIII/** 5, 109–114 (2010)
7. B. Büttgen, T. Oggier, M. Lehmann, CCD/CMOS lock-in pixel for range imaging: challenges, limitations and state-of-the-art, in *Proceedings of 1st Range Imaging Research Day* (Zurich, Switzerland, 2005), pp. 21–32
8. J. Canny, A computational approach to edge detection. IEEE Trans. Pattern Anal. Mach. Intell. **8**, 679–714 (1986)
9. F. Chiabrando, R. Chiabrando, D. Piatti, F. Rinaudo, Sensors for 3D imaging: metric evaluation and calibration of a CCD/CMOS Time-Of-Flight camera. Sensors **9**, 10080–10096 (2009)
10. W.J. Christmas, J. Kittler, M. Petrou, Structural matching in computer vision using probabilistic relaxation. PAMI **17**(8), 749–764 (1995)
11. B. Dellen, G. Alenyà, S. Foix, C. Torras, 3D object reconstruction from Swissranger sensors data using a spring-mass model, in *Proceedings 4th International Conference Computer Vision Theory and Application*, vol. 2 (Lisbon, Portugal 2009), pp. 368–372
12. S. Foix, G. Alenyà, J. Andrade-Cetto, C. Torras Object modeling using a TOF camera under an uncertainty reduction approach, in *Proceedings IEEE International Conference* on *Robotics* and *Automation* (Anchorage, AK, 2010), pp. 1306–1312

13. S. Gehrke, K. Morin, M. Downey, N. Boehrer, T. Fuchs, Semi-global matching: an alternative to LiDAR for DSM generation? International archives of photogrammetry and remote sensing and spatial information sciences—canadian geomatics conference XXXVIII(1), (2010)
14. V.H. Hiep, R. Keriven, P. Labatut, J.-P. Pons, Towards high resolution multi-view stereo, Proceedings computer vision and pattern recognition, pp. 1430–1437 (2009)
15. T. Kahlmann, F. Remondino, H. Ingensand, Calibration for increased accuracy of the range imaging camera Swiss ranger, Int. Arch. Photogram. Remote Sens. Spat. Inf. Sci. **XXXVI**, 136–141 (2006)
16. T. Kahlmann, H. Ingensand, Calibration and development for increased accuracy of 3D range imaging cameras. J. Appl. Geodesy **2**(1), 1–11 (2008)
17. W. Karel, S. Ghuffar, N. Pfeifer, Quantifying the distortion of distance observations caused by scattering in Time-Of-Flight range cameras. Int. Arch. Photogram. Remote Sens. Spat. Inf. Sci. **38**(5), 316–321 (2010)
18. T. Kavli, T. Kirkhus, J. Thielmann, B. Jagielski, Modeling and compensating measurement errors caused by scattering Time–Of-Flight cameras. Proc. SPIE 7066, 706604-1 - 706604-10 (2008)
19. R. Lange, Time-Of-Flight range imaging with a custom solid-state image sensor. Proc. SPIE **3823** 180–191 (1999)
20. R. Lange, P. Seitz, Solid-state Time-Of-Flight range camera. IEEE J. Quantum Electron. **37**(3), 390–397 (2001)
21. R. Lewis, C. Sequin, Generation of 3D building models from 2D architectural plans. Comput. Aided Des. **30**(10), 765–779 (1998)
22. D.D. Lichti, , C. Kim, S. Jamtsho, An integrated bundle adjustment approach to range-camera geometric self-calibration. ISPRS J. Photogramm. Remote Sens. **65**(4), 360–368 (2010).
23. D. Lichti, Self-Calibration of a 3D Range Camera 2008. Int. Arch. Photogram. Remote Sens. Spat. Inf. Sci. **XXXVII**, 927–932 (2008)
24. D.D. Lichti, C. Kim, A comparison of three geometric self-calibration methods for range cameras. J. Remote Sens. **3**, 1014–1028 (2011)
25. M. Lindner, A. Kolb, Lateral and depth calibration of PMD-distance sensors. Proc. ISVC **2** 524–533 (2006)
26. A. Lingua, F. Rinaudo, Aerial triangulation data acquisition using a low cost digital photogrammetric system. Int. Arch. Photogram. Remote Sens **XXXIII**/B2, 449–454, ISSN: 0256-184 (2000)
27. A. Lingua, D. Marenchino, F. Nex, Performance analysis of the SIFT operator for automatic feature extraction and matching in photogrammetric applications. Sensors **9**(5), 3745–3766, ISSN: 1424-8220, (2009) doi: 10.3390/s90503745
28. M. Lo Turco, M. Sanna, Digital modelling for architectural reconstruction. The case study of the Chiesa Confraternita della Misericordia in Turin. Int. Arch. Photogram. Remote Sens. Spat. Inf. Sci. **XXXVIII**-3/W8, 101–106 (2009)
29. J. Mure-Dubois, H. Hugli, Optimized scattering compensation for Time-Of-Flight camera. Proceeding two- and three-dimensional methods for inspection and metrology V, SPIE **6762**, 67620H-1–67620H-10 (2007)
30. F. Nex, Multi-image matching and LiDAR data new integration approach. Ph.D. Thesis, Politiecnico di Torino, Torino (2010)
31. F. Nex, F. Rinaudo, New integration approach of photogrammetric and LIDAR techninques for architectural surveys. Int. Arch. Photogram. Remote Sens. Spat. Inf. Sci. **XXXVIII**-3/W8, 12–17, ISSN: 0256-1840, (2009)
32. D. Piatti, Time-Of-Flight cameras: tests, calibration and multi-frame registration for automatic 3D object reconstruction. Ph.D. Thesis, Politecnico di Torino, Torino (2011)
33. M. Pierrot-Deseilligny, I. Cléry, APERO, an open source bundle adjustment software for automatic calibration and orientation of a set of images. Int. Arch. Photogram. Remote Sens. **XXXVIII-**5/W16, (2011)

34. H. Rapp, M. Frank, F.A. Hamprecht, B. Jähne, A theoretical and experimental investigation of the systematic errors and statistical uncertainties of Time-Of-Flight-cameras. IJISTA **2008**(5), 402–413 (2008)
35. F. Remondino, S. El-Hakim, Image-based 3D modeling: a review. Photogram. Rec. **21**(115), 269–291 (2006)
36. F. Remondino, S. El-Hakim, A. Gruen, L. Zhang, Turning images into 3D models. IEEE Signal Process. Mag. **25**(4), 55–64 (2008)
37. P.J. Rousseeuw, A.M. Leroy, *Robust regression and outlier detection; Wiley series in probability and mathematical statistics* (Wiley, New York, 1987)
38. M. Scherer, The 3D-TOF-camera as an innovative and low-cost tool for recording, surveying and visualization a short draft and some first experiences. Int. Arch. Photogram. Remote Sens Spat Inf Sci **XXXVIII**-3/W8, (2009)
39. O. Steiger, J. Felder, S. Weiss, Calibration of Time-Of-Flight range imaging cameras, in *Proceedings of the 15th IEEE ICIP*, San Diego, CA, USA, 1968–1971 (2008)
40. R. Wallis, An approach to the space variant restoration and enhancement of images, In: Proceedings of Symposium on Current Mathematical Problems in Image Science. Monterey CA, USA, 329–340 (1976)
41. C.A. Weyer, K. Bae, K. Lim, D. Lichti, Extensive metric performance evaluation of a 3D range camera. Int. Soc. Photogram. Remote Sens. **XXXVII**, 939–944 (2008)
42. L. Zhang, Automatic digital surface model (DSM) generation from linear array images. Thesis Diss. ETH No. 16078, Technische Wissenschaften ETH Zurich, IGPMitteilung N. 90 (2005)

Indoor Positioning and Navigation Using Time-Of-Flight Cameras

Tobias K. Kohoutek, David Droeschel, Rainer Mautz and Sven Behnke

1 Introduction

The development of indoor positioning techniques is booming. There is a significant demand for systems that have the capability to determine the 3D location of objects in indoor environments for automation, warehousing and logistics. Tracking of people in indoor environments has become vital during firefighting operations, in hospitals and in homes for vulnerable people and particularly for vision impaired or elderly people [1]. Along with the implementation of innovative methods to increase the capabilities in indoor positioning, the number of application areas is growing significantly. The search for alternative indoor positioning methods is driven by the poor performance of Global Navigation Satellite Systems (GNSS) within buildings. Geodetic methods such as total stations or rotational lasers can reach millimeter level of accuracy, but are not economical for most applications. In recent years, network based methods which obtain range or time of flight measurements between network nodes have become a significant alternative for applications at decimeter level accuracy. The measured distances can be used to determine the 3D position of a device by spatial resection or multilateration. Wireless devices enjoy widespread use in numerous diverse applications including sensor networks, which can consist of countless embedded devices, equipped with sensing capabilities, deployed in all environments and organizing themselves in an ad-hoc fashion [2]. However, knowing the correct positions of network nodes and their deployment is an essential precondition. There are a large number of alternative positioning technologies

T. K. Kohoutek (✉) · R. Mautz
ETH Zurich—Institute for Geodesy and Photogrammetry,
Wolfgang-Pauli-Str. 15 8093 Zurich, Switzerland
e-mail: kohoutek@geod.baug.ethz.ch

D. Droeschel · S. Behnke
Rheinische Friedrich-Wilhelms-Universität Bonn—Institute for Informatics VI,
Friedrich-Ebert-Allee 144 53113 Bonn, Germany
e-mail: droeschel@ais.uni-bonn.de

F. Remondino and D. Stoppa (eds.), *TOF Range-Imaging Cameras*,
DOI: 10.1007/978-3-642-27523-4_8, © Springer-Verlag Berlin Heidelberg 2013

(Fig. 1) that cannot be detailed within the scope of this paper. An exhaustive overview of current indoor position technology is given in [3]. Further focus will be on optical methods.

Optical indoor positioning systems can be categorized into static sensors that locate moving objects in the images and ego-motion systems whose main purpose is the position determination of a mobile sensor (i.e. the camera) [4]. Some optical system architectures do not require the deployment of any physical reference infrastructure inside buildings, which can be a requirement for a widespread implementation.

This article investigates the use of Time-Of-Flight (TOF) cameras for ego-motion determination in indoor environments. TOF cameras are suitable sensors for simultaneous localization and mapping (SLAM), e.g. onboard of autonomous Unmanned Vehicle Systems (UVS), or the detection and localization of objects in indoor environments. They are an attractive type of sensor for indoor mapping applications owing to their high acquisition rate collecting three-dimensional (3D) data. TOF cameras consist of compact, solid-state sensors that provide depth and reflectance measurements at high frame rates of up to 50 Hz independent from surrounding light.

The approximate 3D position accuracy for objects seen by the used MESA® TOF camera SwissRanger SR-4000 (in terms of a 1-σ standard deviation) is 1 cm

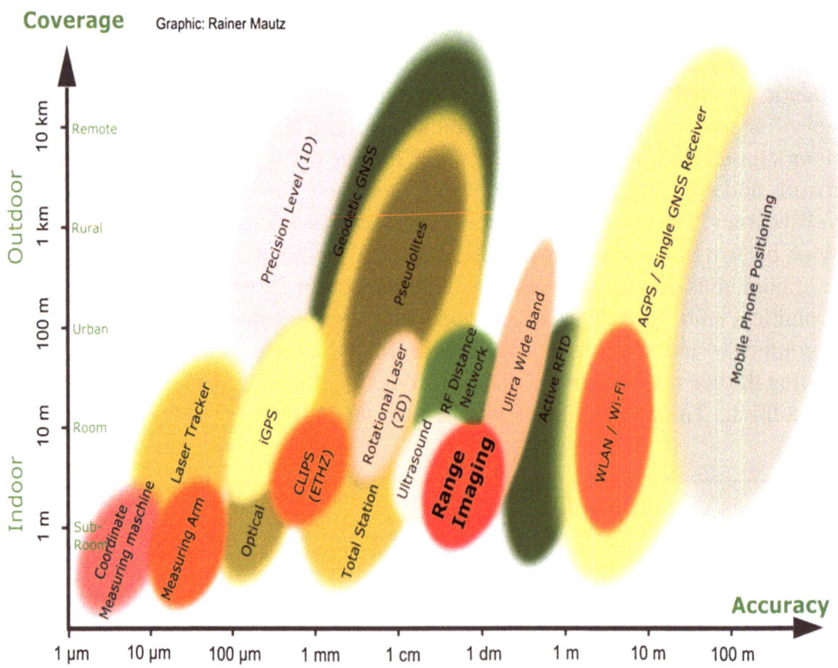

Fig. 1 Today's positioning systems in dependence to accuracy and coverage [1]

for distances of up to 5 m and 1 dm for distances up to 15 m. Such a level of accuracy is sufficient for some indoor applications, e.g. collision avoidance. Currently, ranges larger than 15 m and accuracies better than 1 cm are not applicable to TOF cameras. In these cases 3D laser scanners or stereo/multiple camera systems need to be used instead. As a drawback of two-dimensional (2D) cameras, the prerequisite for multiple views induces a high computational load since point correspondences between at least two images from different perspectives have to be determined. In addition, distances to structureless surfaces cannot be measured, because the correspondence problem [5] cannot be solved. Furthermore, passive 2D vision suffers from shadowing effects and sensitivity to changes in illumination. The use of 3D laser range finders [6] that actively illuminate the scene can avoid these issues but needs mechanical moving parts and have high power consumption as well as a low frame rate due to sequential point acquisition.

Our procedure is as follows. Image features, e.g. edges, corners or flat surfaces are detected based on reflectance data for object recognition in the indoor environment. In Sect. 2 we will show how the indoor positioning with the TOF camera can be realized. As a novelty, the proposed method combines absolute and relative orientation of a TOF camera without the need for dedicated markers or any other locally deployed infrastructure. This can be achieved, because in comparison to other methods range imaging directly provides 3D point clouds that are compared with a spatio-semantic 3D geoinformation model offered by the City Geographic Markup Language (CityGML) that supports any coordinate system and enables the missing link between the indoor and outdoor space. As higher the level of semantic information as more accurate is the geometrical integration. The entrance door of a building for example is always connected to a walkable surface. The camera motion is estimated based on depth data and will be explained within the mapping process in Sect. 3. Collision avoidance becomes important if the navigation path is unknown. Section 4 will show that TOF cameras are ideally suited for that task. A welcome side effect of our approach is the generation of 3D building models from the observed point cloud.

2 Positioning Inside the Room Based on a CityGML Model

The standard CityGML [7] defines a data model and an XML data format for 3D city and topography models. CityGML defines several Levels of Detail (LoD) with the highest LoD 4 having the capability for modeling the interior of buildings. In particular for the purpose of indoor modeling, the semantic model provides an object class ‘Room’ that can capture semantic data [8], including attributes for the intended and current use of the room such as ‘Living Room’ or ‘Office’. An object of the class ‘Room’ can be associated with its geometry in two different ways. In one way the outer shell of a room can be defined by establishing a link to a geometric object of type *Solid* or *MultiSurface* (both types are defined by the GML 3.1.1 specification [9]). Alternatively, the outer shell can be decomposed into semantic objects of the types *InteriorWallSurface*, *CeilingSurface* and *FloorSurface*, which are referred to geometric objects of type *MultiSurface*. Openings in the outer shell of a room can be modeled with the object classes 'Window' and 'Door' that can belong to one or two *InteriorWallSurfaces*. This data structure can be used to express topological relationships between rooms.

The semantic object class *IntBuildingInstallation* can be used to model permanent fixed objects belonging to a room e.g. radiators, columns and beams. In order to model the mobile components of a room such as desks and chairs, the object class *BuildingFurniture* can be used. *IntBuildingInstallation* and *BuildingFurniture* provide the attribute class for semantic description of the objects (Fig. 2). The geometry of these fixed installed objects can be defined by the standard GML 3.1.1. So-called implicit geometries are used to model simplified shapes of the movable objects in a room. Hereby the shape of an object is stored only once in the library even if multiple objects of the same shape are present (e.g. pieces of furniture). The shapes could be obtained directly from the 3D CAD drawings of pieces of furniture in the manufacturer's catalog. For each occurrence of such an object, only the local coordinates of an insertion point and the object's orientation are stored. The orientation parameters are linked to the geometry that has become an object of CityGML.

Nowadays, Building Information Models (BIMs) are created within the planning and construction phase of a building [10]. The acquisition of BIMs for already existing buildings requires manual measurements using total stations, terrestrial laser scanners or photogrammetric techniques. Figure 3 illustrates semantic classification of CityGML exemplified with an indoor model of a room that has been obtained by total station survey.

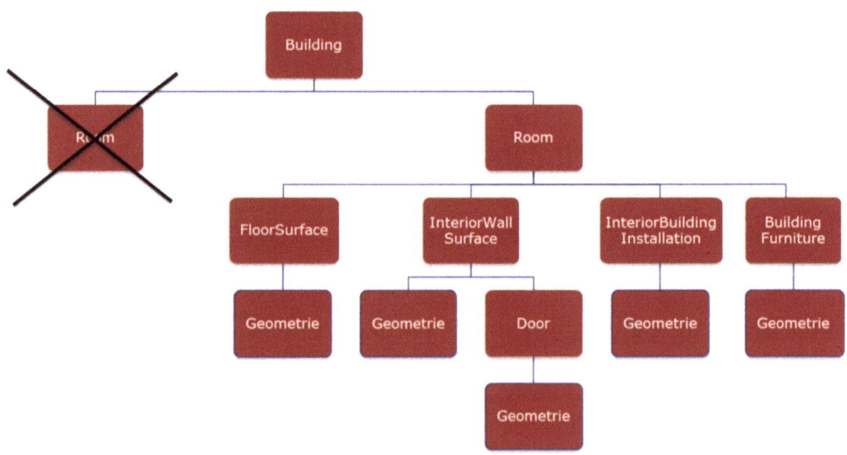

Fig. 2 Decision tree for room identification

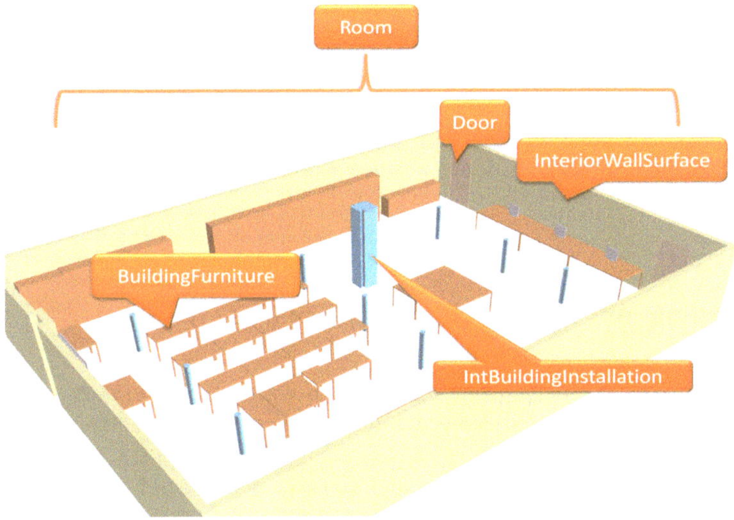

Fig. 3 ETH Zurich lecture room modeled in CityGML [11]

2.1 Room Identification Through Object Detection

Object detection is the key challenge for the correct identification of the room where the sensor is located. The detection of objects can be achieved by exploiting the amplitude image. In order to identify objects such as chairs, tables, etc., the known or "learned" primitives, features and image templates that have previously stored in the database are matched with the current image. The detected object properties such as the size, geometry or quantity of a certain object are the main criteria for the comparison with the database. This way, the unknown camera position can be limited to a small number of possible rooms in the building. The room can be identified uniquely by detecting its distinct properties, e.g. position of installations. After a successful identification additional semantic and geographic information can be extracted from the 3D geo database.

2.2 Accurate Positioning Using Distance Measurements

This step compares and transforms the in real time acquired Cartesian 3D coordinates of the objects into the reference coordinate system of the database. All room and object models in the CityGML database are saved as Virtual Reality Modeling Language (VRML) files. Suitable reference points for the transformation (with 6 degrees of freedom) are the corners of the room, vertices of doors, windows and other fixed installations. The accuracy of the objects in CityGML should be at centimeter level and should lead to position determination of the camera with centimeter-accuracy using a least squares adjustment with a redundant number of reference points to determine the 3D camera position. One requirement for the camera is that its interior orientation has been determined previously. The exterior camera orientation (3 translations and 3 rotations) is determined by a Cartesian 3D coordinate transformation with 3 shift and 3 rotational parameters. There is no need to estimate a scale parameter, since calibrated TOF cameras measure the absolute distance.

3 Mapping and Ego-Motion Estimation

Dense depth measurements from TOF cameras enable the generation of 3D maps of the camera's environment. However, the accuracy of measurements in unknown scenes varies considerably, due to error effects inherent to their functional principle. Therefore, a set of preprocessing steps to discard and correct noisy and erroneous measurements need to be applied in order to achieve accuracy according to the specification.

3.1 Sensor Data Processing

First, mixed pixels at so-called jump edges are filtered out. Mixed pixels are a result of false measurements that occur when the signal from the TOF camera hits an edge of an object. Then, the signal is partially reflected at the foreground, but also at the background. Both signal parts arrive at the same CCD element. The true distance changes suddenly at the object border, but the values of the mixed pixels consist of an average between the foreground and background distance. In the point cloud, these pixels appear as single unconnected points that seem to float in the air and that do not belong to any object. This is also a common problem in terrestrial laser scanning. Jump edges are filtered by local neighborhood relations comparing the opposing angles of a point p_i and its eight neighbors $p_{i,n}$, [12]. From a set of 3D points $P = \{ p_i \in R^3 | i = 1, \ldots, N_p \}$, jump edges are detected by comparing opposing angles $\theta_{i,n}$ of the triangle spanned by the focal point $f = 0$ and its eight neighbors $P_n = \{ p_{i,n} \in R^3 | i = 1, \ldots N_p : n = 1, \ldots, 8 \}$ and filtered with a threshold θ_{th}:

$$\theta_i = \max \arcsin \left(\frac{\|p_{i,n}\|}{\|p_{i,n} - p_i\|} \sin \varphi \right), \tag{1}$$

$$J = \{ p_i | \theta_i > \theta_{th} \}, \tag{2}$$

where φ is the apex angle between two neighboring pixels. Since the jump edge filter is sensitive to noise, a median filter is applied to the distance image beforehand. Besides mixed pixels, measurements with low amplitude are neglected since the accuracy of distance measurements is dependent on the amount of light returning to the sensor.

TOF cameras gain depth information by measuring the phase shift between emitted and reflected light, which is proportional to the object's distance modulo the wavelength of the modulation frequency. As a consequence, a distance ambiguity arises: measurements beyond the sensor's wavelength are wrapped back causing artifacts and spurious distance measurements. Wrapped distance measurements can be corrected by identifying a number of so-called phase jumps in the distance image, i.e., the relative wrappings between every pair of neighboring measurements. Droeschel et al. proposed attempt a probabilistic approach that detects discontinuities in the depth image to infer phase jumps using a graphical model [13]. Every node in the graphical model is connected to adjacent image pixels and represents the probability of a phase jump between them. Belief propagation is used to detect the locations of the phase jumps which are integrated into the depth image by carrying out the respective projections, thereby correcting the erroneously wrapped distance measurements. The application of phase unwrapping for an indoor scene is shown in Fig. 4.

3.2 Mapping and Ego-Motion Estimation

To estimate the camera's motion between two consecutive frames, image features in the reflectance image of the TOF camera are extracted to determine point correspondences between the frames. To detect image features, the Scale Invariant

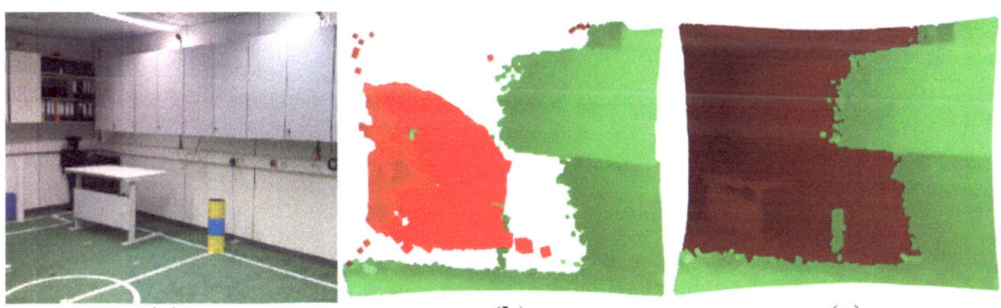

(a) (b) (c)

Fig. 4 Phase unwrapping of an indoor scene. **a** Image of the scene. **b** and **c** 3D point clouds that have been generated based on the camera's depth image. *Color* of the points indicates the result of the algorithm; wrapped measurements are shown in *red*. *Brightness* encodes distance to the camera center. **b** Point cloud without unwrapping. Measured distances beyond the sensor's non-ambiguity range are wrapped into it, which results in artifacts between distances of 0 and 3 meters. **c** Unwrapped depth image

Feature Transform (SIFT) [14] is used. SIFT features are invariant in rotation and scale and are robust against noise and illumination changes.

In order to estimate the camera motion between two frames, the features of one frame are matched against the features of the other frame. The best match is the nearest neighbor in a 128-dimensional keypoint descriptor space. To determine the nearest neighbor, the Euclidean distance is used. In order to measure the quality of a match, a distance ratio between the nearest neighbor and the second-nearest neighbor is considered. If both are too similar, the match is rejected. Hence, only features that are unambiguous in the descriptor space are considered as matches.

Figure 5a and b show the reflectance image of two consecutive frames with detected features. Figure 5c shows the matching result of the two images. Each match constitutes a point correspondence between two frames. By knowing the depth of every pixel, a point correspondence in 3D is known.

The set of points from the current frame is called the data set, and the set of corresponding points in the previous frame is called the model set. The scene is translated and rotated by the sensor's ego motion. Thus, the sensor's ego motion can be deduced by finding the best transformation that maps the data set to the model set. A common approach for estimating a rigid transformation uses a closed form solution for estimating the 3×3 rotation matrix R and the translation vector t, which is based on singular value decomposition (SVD) [15]. The distances between corresponding points, after applying the estimated transformation are used to compute the root mean square error (RMSE) which is often used in range registration to evaluate the scene-to-model consistency. It can be seen as a measure for the quality of the match: if the RMSE is significantly high, the scene-to-model registration cannot be consistent. On the other hand, a low RMSE does not imply a consistent scene-to-model registration, since it also depends on the number and distribution of the point correspondences.

With the estimated ego motion between consecutive frames, accumulating 3D points of every frame generates a point-based map. A resulting map is shown in Fig. 6.

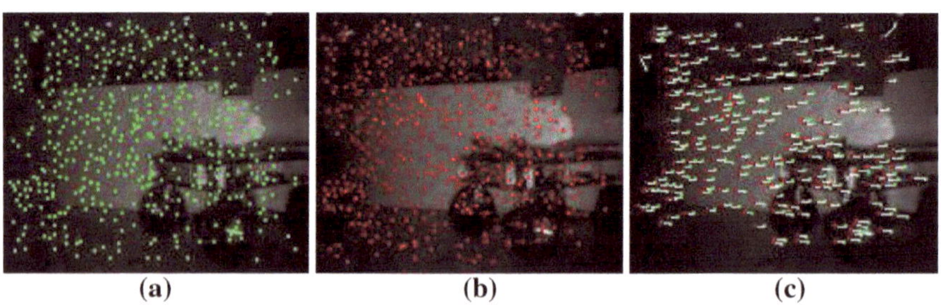

Fig. 5 SIFT feature extraction and matching applied on two consecutive camera frames on a TOF reflectance image. The numbers of detected features are 475 (**a**) and 458 (**b**). **c** Matching result: 245 features from image (**a**) are matched to features from image (**b**). *White lines* indicate feature displacement

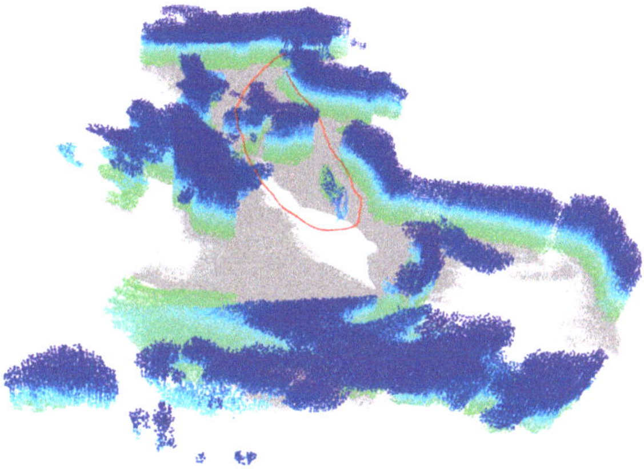

Fig. 6 The resulting 3D map based on the estimated trajectory (*red*). The *colors* of the points correspond to the distance of the point from the ground plane

4 3D Collision Avoidance

If the navigation path is unknown in dynamic environments, collision avoidance becomes important. TOF cameras are ideally suited for collision avoidance since they measure distances to surfaces at high frame rates.

A typical example of a point cloud taken in an indoor environment is shown in Fig. 7a. This point cloud can be used to build a so-called height image as shown in Fig. 7b. A point $p_{i,\ j}$ is classified as belonging to an obstacle if

$$(W_{max} - W_{min}) > H, \tag{3}$$

where W_{max} and W_{min} are the maximum and minimum height values from a local window W, spanned by the eight-connected neighborhood around $p_{i,\ j}$. The Threshold H thereby corresponds to the minimum tolerable height of an obstacle.

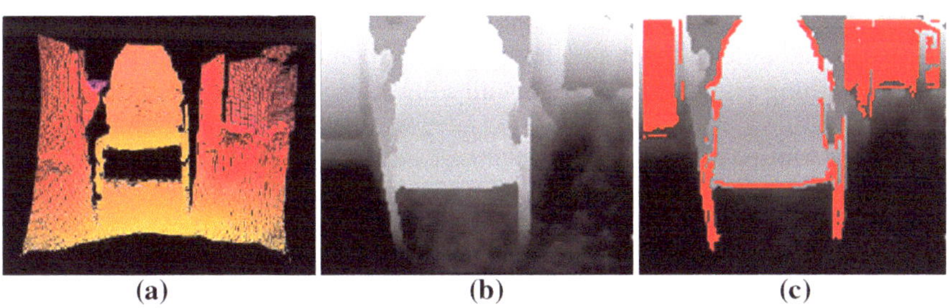

Fig. 7 **a** 3D Point cloud of an exemplary scene. The *color* of the points corresponds to the distance, *brighter color* relates to shorter distances and *darker color* to farther distances. **b** The generated height image. The grayscale value of every pixel corresponds to the z-coordinate of the respective point in the point cloud. **c** The resulting obstacle points (*red*)

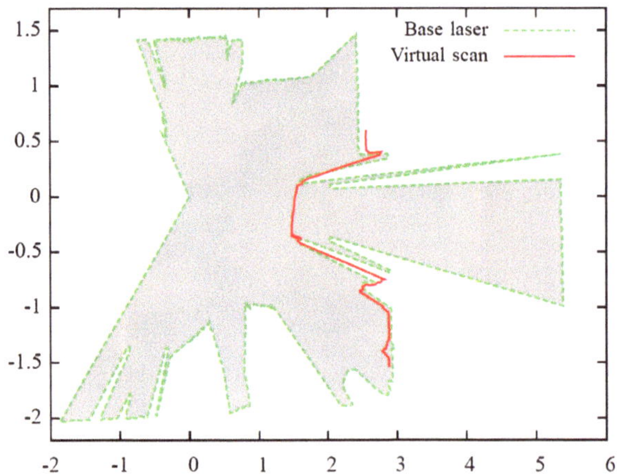

Fig. 8 The resulting virtual scan of the scene is compared with the scan from the laser range finder. The dashed *green line* illustrates the base laser scan. The *red line* illustrates the virtual laser scan. The chair shows only a few points in the base laser scan since only the legs of the chair are in the scan plane, whereas the virtual scan outlines the contour of the chair

It needs to be chosen appropriately since it should not be smaller than the sensor's measurement accuracy. Due to evaluating a point's local neighborhood, floor points are inherently not considered as obstacles. Points classified as belonging to obstacles are shown in Fig. 7c.

The resulting obstacle points are used to extract a 2D virtual scan similar to an obstacle map by (1) projecting the 3D data into the xy-plane and (2) extracting relevant information.

The number of range readings in the virtual scan as well as its apex angle and resolution correspond to the acquired 3D data. For the SR4000, the number of range readings is 176, which is the number of columns in the image array. The apex angle and the angular resolution are 43 and 0.23°, which correspond to the camera's horizontal apex angle and resolution. For every column of the TOF camera's distance image, the obstacle point with the shortest Euclidean distance to the robot is chosen. This distance constitutes the range reading in the scan. If no obstacle point is detected in a column, the scan point is marked invalid.

The resulting virtual scan is fused with a 2D laser range scan obtained at 30 cm height yielding a common obstacle map modeling the closest objects in both sensors. The obstacle map from the 2D laser range finder and the TOF camera for the aforementioned example scenario is visualized in Fig. 8. By fusing the information of both sensors, the robot possesses correct information about traversable free space (light gray) in its immediate vicinity.

5 Conclusions and Outlook

Efficient and precise position determination of a TOF camera is possible based on kinematic object acquisition in form of 3D Cartesian coordinates. The absolute position of the camera can be obtained by a transformation from the camera coordinate system into the reference coordinate system, i.e. the coordinate system of the spatio-semantic 3D model. Positions of detected objects are reported in respect to the coordinate system of the 3D model. The described mapping approach can also be used for data acquisition of such 3D building models. The advantage of such models is the use of the VRML file text format allowing data compression for the purpose of quick Internet transfer and maintenance of a small-sized database. We conclude that rooms can be identified by detection of unique objects in images or point clouds. Such method is to be implemented in further research based on a data set, which includes multiple rooms.

Due to their measurement of a volume at high frame rates, TOF cameras are well suited for applications where either the sensor or the measured objects move quickly, such as 3D obstacle avoidance, measured or gesture recognition [16].

Difficulties of the proposed method arise from TOF cameras suffering from a set of error sources that hamper the goal of infrastructure-free indoor positioning. Current range imaging sensors are able to measure distances unambiguously between 5–15 m at an accuracy level of centimeters. Until now so-called mixed pixels posed a problem in the literature. Filtering methods presented in Sect. 3 could solve this problem.

Acknowledgments The support of Andreas Donaubauer, Andreas Schmidt, Dirk Holz and Stefan May is gracefully acknowledged.

References

1. T.K. Kohoutek, R. Mautz, A. Donaubauer, Real-time indoor positioning using range imaging sensors, in *Proceedings of SPIE Photonics Europe, Real-Time Image and Video Processing*, vol. 7724 (SPIE, 2010), p. 77240K. doi:10.1117/12.853688 (CCC code: 0277-786X/10/$18)
2. R. Mautz, W.Y. Ochieng, Indoor positioning using wireless distances between motes, in *Proceedings of TimeNav'07 / IEEE International Frequency Control Symposium*, (Geneva, Switzerland, 29 May–1 June 2007), pp. 1530–1541
3. R. Mautz, Indoor positioning technologies, in *A habilitation thesis submitted to ETH Zurich for the Venia Legendi in Positioning and Engineering Geodesy*, (ETH, Zurich, 2012)
4. R. Mautz, S. Tilch, Survey of optical indoor positioning systems, in *IEEE Xplore Proceedings of the 2011 International Conference on Indoor Positioning and Indoor Navigation (IPIN)*, (Guimarães, Portugal, 21–23 Sept 2011)

5. B. Julesz, Binocular depth perception of computer-generated patterns. Bell System Tech. **39**(5), 1125–1161 (1960)
6. C. Keßler, C. Ascher, G.F. Trommer, Multi-sensor indoor navigation system with vision- and laser-based localisation and mapping capabilities. Eur. J. Navig. **9**(3), 4–11 (2011)
7. G. Gröger, T.H. Kolbe, A. Czerwinski, C. Nagel: *OpenGIS® City Geography Markup Language (CityGML) Encoding Standard Version 1.0.0*, (International OGC Standard. Open GeospatialConsortium, Doc. No. 08-007r1, 2008)
8. G. Gröger, T.H. Kolbe, A. Czerwinski: Candidate *OpenGIS® CityGML Implementation Specification (City Geography Markup Language) Version 0.4.0*, (International OGC Standard. Open Geospatial Consortium, Doc. No. 07-062, 2007)
9. S. Cox, P. Daisey, R. Lake, C. Portele and A. Whiteside: *OpenGIS® Geography Markup Language (GML) Implementation Specification Version 3.1.1*, (International OGC Standard. Open Geospatial Consortium, Doc. No. 03-105r1, 2004)
10. C. Nagel, A. Stadler, T.H. Kolbe, Conceptual requirements for the automatic reconstruction of building information models from uninterpreted 3D models, in *The International Archives of the Photogrammetry, Remote Sensing and Spatial Information Sciences*, vol. 34, Part XXX (2009)
11. A. Donaubauer, T.K. Kohoutek, R. Mautz, CityGML als Grundlage für die Indoor Positionierung mittels Range Imaging, in *Proceedings of 15. Münchner Fortbildungsseminar Geoinformationssysteme*, (Munich, Germany, 8–11 March 2010), pp. 168–181
12. S. May, D. Droeschel, D. Holz, S. Fuchs, E. Malis, A. Nüchter and J. Hertzberg, Three-dimensional mapping with Time-Of-Flight cameras, J. Field Robot. **26**(11–12), 934–965 (2009). (*Special Issue on Three-dimensional Mapping, Part 2*)
13. D. Droeschel, D. Holz, S. Behnke, Probabilistic phase unwrapping for Time-Of Flight cameras, in *Proceedings of the joint conference of the 41st International Symposium on Robotics (ISR 2010) and the 6th German Conference on Robotics (ROBOTIK 2010)*, (Munich, Germany, 2010), pp. 318–324
14. D.G. Lowe, Distinctive image features from scale invariant keypoints. Int. J. Comput. Vision **60**, 91–110 (2004)
15. K.S. Arun, T.S. Huang, S.D. Blostein, Least-squares fitting of two 3-d point sets. IEEE Trans. Pattern Anal. Mach. Intell. **9**(5), 698–700 (1987)
16. D. Droeschel, J. Stückler, S. Behnke, Learning to interpret pointing gestures with a Time-Of-Flight camera, in *Proceedings of the 6th ACM/IEEE International Conference on Human-Robot Interaction (HRI)*, (Lausanne, Switzerland, March 2011), pp. 481–488

TOF Cameras and Stereo Systems: Comparison and Data Fusion

Carlo Dal Mutto, Pietro Zanuttigh and Guido M. Cortelazzo

Time-Of-Flight range cameras and stereo vision systems (for simplicity called TOF cameras and stereo systems now on) are both depth acquisition devices capable to collect 3D information of dynamic scenes. In spite they can be used for similar tasks in many applications, it would not be appropriate to view the two systems as alternate or even competitive choices, since their characteristics and actual capability are markedly different. Indeed synergically combining together TOF cameras and stereo systems is a rather intriguing and useful option. This chapter firstly compares TOF range cameras and stereo vision systems, and then addresses the problem of fusing the data produced by the two systems. Because of the many aspects involved, the comparison is all but straightforward and could be certainly organized in different ways. The proposed one represents a systematic approach.

This chapter is organized in three sections. The first section reviews stereo systems (TOF technology has already been covered in detail in this book). The basic ingredients of a stereo system essentially are a pair of cameras and a specific implementation of the computational stereopsis procedure (for simplicity just called stereo algorithm from now on). The basic geometrical properties of the acquisition systems are first introduced in order to analyse their influence on the final quality of the stereo measurements. Some classical stereo algorithms are subsequently introduced. Because of the great number of stereo algorithms the selected subset represents the methods currently considered of greatest interest.

The second section of the chapter compares TOF cameras and the stereo measurements. Among other things TOF depth data are generally more robust than the ones from stereo systems, particularly in case of texture-less scenes. In case one needs high spatial resolution, stereo systems are probably the best choice.

The third and last part of the chapter covers the various methods proposed in literature for the fusion of TOF and stereo data and with the double purpose of both

C. D. Mutto (✉) · P. Zanuttigh · G. M. Cortelazzo
University of Padua, Padua, Italy
e-mail: dalmutto@dei.unipd.it

F. Remondino and D. Stoppa (eds.), *TOF Range-Imaging Cameras*,
DOI: 10.1007/978-3-642-27523-4_9,

reviewing the literature and of offering our own view on how to improve upon current approaches.

1 Stereo Vision Systems

An image-based stereo system employs images from a pair of standard video-cameras in order to derive a depth map of the scene simultaneously framed by the two cameras. All stereo systems exploit the triangulation measurement principle of the photogrammetric technique [1, 2]: given two cameras pointing towards an object, the difference between the positions of the object in the two acquired images (the so called parallax) is inversely proportional to the distance of the object from the cameras (as later formalized in this chapter). Two examples of commercial stereo vision systems are delivered by Point Gray [3] and TZYX [4].

1.1 Basic Geometry of a Stereo Vision System

The "hardware component" of the stereo system is mainly constituted by a pair of standard video-cameras and optionally by a synchronization circuit rather useful in case of dynamic scenes. The depth information computed by the stereo system is relative to the point of view of one of the two cameras, usually called the *reference* camera, while the other one is usually called *target* camera. In this chapter the reference camera will be the left one (denoted by L) and the target the right one (denoted by R). The acquired images are called either reference or target images depending on the camera acquiring them (Fig. 1).

A stereo acquisition system can exploit either grayscale or colour cameras. Of course colour acquisition systems may run stereo algorithms accounting for colour information, typically more robust and precise than grayscale stereo algorithms

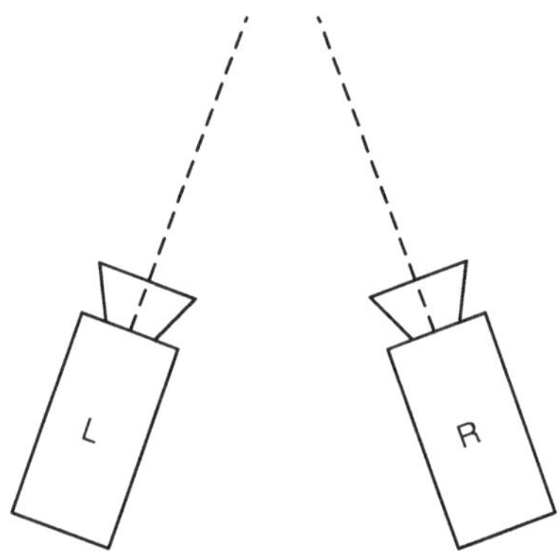

Fig. 1 *Left* (reference) camera *L* and *right* (target) camera *R* with their optical axes

(currently the top ranking algorithms according to the Middlebury evaluation website [5] exploit also colour information). Colour stereo comes at an increase of the computational cost with respect to grayscale stereo. The trade-off between robustness and computational complexity is a fundamental issue for stereo systems.

Since the general idea behind stereo vision is to compare the reference and the target images, it is mandatory that such images refer to the same temporal scene configuration, i.e., the two cameras need to be synchronized. This can be obtained via hardware by a synchronization circuit or via software. In presence of fast motion if the two cameras are not well synchronized, the reference and the target images do not refer to the same position of the objects moving in the scene, resulting therefore in several 3D geometry estimation artefacts. Hardware synchronization is strongly recommended against motion artefacts with fast dynamic scenes. Epipolar geometry clearly shows how the final depth estimation quality directly depends on the geometrical configuration of a stereo system. Unfortunately in this chapter there is no room for epipolar geometry which is however treated in many books about stereo vision and photogrammetry, such as [2, 6, 7].

The acquisition system needs to be characterized in order to estimate the intrinsic and extrinsic parameters of both cameras. Several camera calibration tools are available in the vision community to retrieve both the intrinsic and extrinsic parameters of a camera. In particular, [8, 9] are state-of-the-art open source projects in this field. In this chapter, the vision system is always supposed calibrated and the parameters were computed using the tool available in [9].

Given a calibrated stereo system, it is usual to apply an undistortion and rectification procedure to the images acquired by the two cameras in order to simplify the task of stereo vision algorithms. The procedure, which leads to a so-called rectified vision system, takes as input the images acquired by L and R and performs the following operations:

(1) Correction of the radial and tangential distortion introduced by the camera lenses.
(2) Compensation of the focal length differences between L and R.
(3) Compensation of the differences in the other intrinsic parameters of the L and R cameras.
(4) Compensation of the relative rotation between the two cameras in order to obtain images as if they were acquired by cameras with parallel optical axes orthogonal to the line through the optical centre of L and R.

For details on stereo vision system rectification, the reader is referred to [6, 7, 10].

Some notation used in the rest of the chapter is afterwards introduced. Each one of the two stereo cameras is associated to a standard 3D reference system, cantered at the centre of projection or nodal point of the camera, with z-axis oriented along the camera optical axis, x-axis horizontally oriented and y-axis vertically oriented as shown in Fig. 2. The image acquired by L, after rectification is called rectified reference image, and denoted by I_L. The image acquired by R, after the

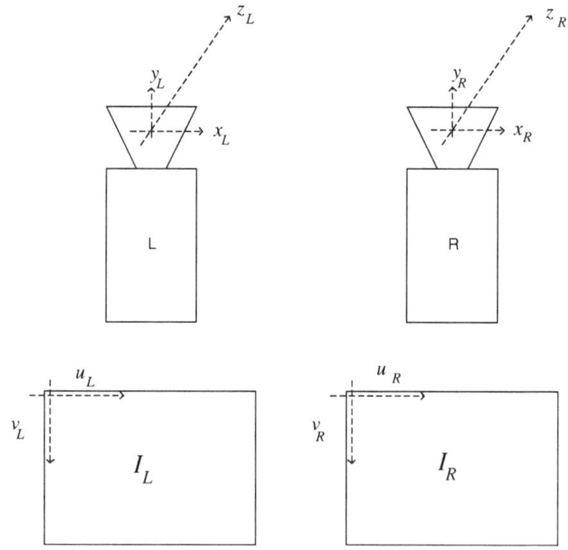

Fig. 2 Acquisition system (camera L and camera R), rectified reference image I_L, rectified target image I_R and their relative 3D and 2D reference systems

rectification process is called rectified target image, and denoted by I_R. The two images I_L and I_R are associated each one to a standard 2D reference system, with horizontal axis u pointing rightward and vertical axis v pointing downward as shown in Fig. 2.

It is worth pointing out that for a rectified stereo system no rotation is assumed between the 3D reference systems associated with L and R as well as no translation along the y and z directions.

A scene point $P = [x, y, z]^T$ with coordinates expressed with respect to the L 3D reference frame, if visible from both cameras, is projected to point $p_L = [u, v]^T$ on I_L with coordinated expressed with respect to the I_L 2D reference system and to point $p_R = [u - d, v]^T$ on I_R with coordinated expressed with respect to the I_R 2D reference system. It can be shown that the difference d between the coordinates of the two 2D points, called disparity, and the depth value z of P are related as:

$$d = \frac{bf}{z} \tag{1}$$

where b is the baseline, i.e., the distance between the nodal points of L and R, and f is the focal length (assumed equal for both rectified cameras). Points p_L and p_R, called conjugate points, share the same vertical coordinate v. One can associate a disparity value to each pixel p_L and obtain an image of disparity values, denoted as I_D and called disparity image or disparity map. From (1) it is clear that high values of d correspond to points close to the cameras, i.e., to points with low z value. Also, since d is generally quantized and there is an inverse relationship between z and d, the accuracy of the stereo vision systems does not decrease linearly, but quadratically with respect to z according to the following equation [4]:

$$\Delta z = \frac{z^2}{fb} \Delta d \tag{2}$$

where Δd is the disparity quantization step and Δz the depth quantization step. As the image pair has been rectified, there are no negative values of d are valid and $d = 0$ corresponds to points with depth value $z = \infty$. It is customary to limit the range of the values that d may take on the basis of geometrical considerations. If the minimum and the maximum depth values (respectively z_{MIN} and z_{MAX}) of the scene are known, the disparity excursion, can be confined to $d \in [d_{MIN}, d_{MAX}]$, with $d_{MIN} = \frac{bf}{z_{MAX}}$ and $d_{MAX} = \frac{bf}{z_{MIN}}$.

1.2 Stereo Vision Algorithms

It has been shown in the previous section that for a rectified stereo system, the value of the depth distribution z of the scene points $P = [x, y, z]^T$ visible from both cameras can be obtained by (1) from the estimation of disparity distribution d between all the pairs of conjugate points $p_L = [u, v]^T \in I_L$ and $p_R = [u - d, v]^T \in I_R$. Hence the information about the depth distribution of a scene is coded by the disparity image I_D, which is a typical intermediate output of stereo algorithms. The computation of the depth distribution of the framed scene is typically called computational stereopsis [6] and encompasses two steps, the first is a point matching procedure corresponding to a linear search meant to detect conjugate points along each horizontal line of I_L, row by row and the second is the computation of the depth distribution z from the disparity image I_D by (1). The point matching is a rather critical step since wrong matches inevitably lead to wrong scene depth estimates. Manual point selection ensures correct points association at the expenses of considerable labour. Computer vision exclusively focuses on automatic point matching procedures with the advantage that they can be made by machines without any human intervention but with the risk of incorrect matches. Stereo matching can be performed in many ways, essentially trading speed against robustness, and it is a distinctive element differentiating the various stereo methods. A wide class of stereo algorithms, called local methods, exploits local similarity in order to detect, given p_L in I_L, the point p_L on the corresponding line of I_R with neighbourhood most similar to that of p_L (of course similarity can be defined in many ways). Other algorithms, called global methods, adopt a global model of the scene, by implicitly or explicitly imposing constraints on the overall scene depth configuration. Semi-global methods use scene models imposing constraints only on parts of the scene depth. The following subsections review three examples of these methods currently in great consideration and usage: namely, the most classical local algorithm, i.e., Fixed Window (FW); a widely adopted global algorithm, i.e., Loopy Belief Propagation (LBP); and Semi Global Matching (SGM), a state of the art semi-global algorithm. The literature on stereo vision algorithms is extremely vast. The ones presented in this chapter are widely used and implemented in the OpenCV computer vision library [9]. A more detailed analysis of stereo vision methods can be found in [6, 11]. The Middlebury benchmark website [5], reporting and updating the performance ranking of the latest stereo vision algorithms, is an important source of information.

1.3 Local Stereo Algorithms

The FW algorithm is a classical local stereo algorithm still widely used in practical implementations for its simplicity. For each pixel $p_L = (u, v) \in I_L$ its conjugate $p_R = (u - d^*, v) \in I_R$ (and equivalently its disparity value d^*) is computed as follows:

(1) A squared (or rectangular) window W_L is centred around p_L, and other windows of the same size W_R^i are cantered around each candidate conjugate point $p_R^i(u - i, v), i = 1, \ldots, d_{MAX} - d_{MIN}$ as shown in Fig. 3.
(2) The cost c_i of matching p_L against each one of the candidate conjugate points p_R^i is computed by comparing I_L on W_L and I_R on each W_R^i. An example of such a costs and type of comparisons is the Sum of Absolute Differences (SAD), i.e., $c_i = \frac{1}{|W_L|} \sum_{p \in W_L, q \in W_R^i} |I_L(p) - I_R(q)|$, where $|W_L|$ is the number of pixels in W_L. Clearly many other different measures could be used in this task, e.g., the correlation, the sum of squared differences [11] or more complex measures such as Adaptive Least Squares Correlation [12].
(3) Pixel p_R^i corresponding to the minimum matching cost c_i is selected as conjugate of p_L, and $d^* = d_i$.

Such a local method considers a single pixel of I_L at the time, it adopts a Winner-Takes-All (WTA) strategy for disparity optimization, and it does not explicitly impose any model on the depth distributions. Like most local approaches, cost aggregation within fronto-parallel windows implicitly assumes the same disparity for all the points within the window. This is clearly not true if the window includes a scene depth discontinuity. Indeed FW is well known not to perform well across depth discontinuities. Moreover, as most local algorithms, FW performs poorly in texture-less regions. Nevertheless, since incremental calculation

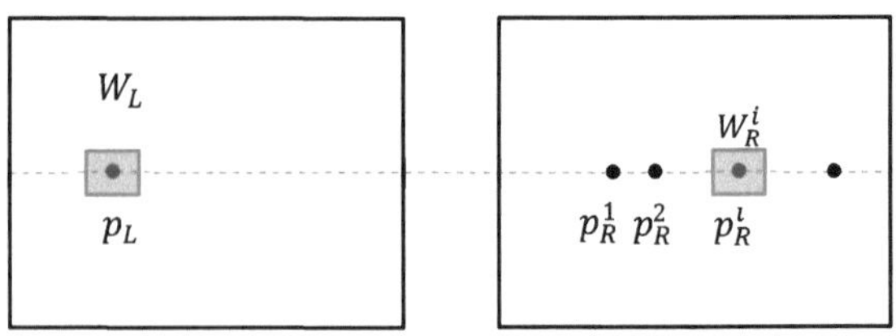

Fig. 3 Fixed Window (FW) stereo algorithm

schemes, e.g. [13, 14], can make FW very fast, it is widely used in practical applications despite its notable limitations. The larger the window size the better the robustness against image noise and low texture situations, at the expense of the precision in presence of discontinuities.

Evolutions of FW focus on the shape of the coupling window [15], on the usage of multiple coupling windows for a single pair of candidate conjugate points [16], on weighting the contribution of the different pixels within a window according to suitable weights, given for instance by a bilateral filter [17] or derived from segmented versions of I_L and I_R [18]. These modifications of the classical fixed window strategy improve its performance, especially in presence of depth discontinuities, but significantly increase computation/execution time.

An interesting variant of FW applies the SAD strategy to colour images I_R and I_T (assumed available) by separately treating their colour channels.

1.4 Global Stereo Algorithms

While local stereo algorithms estimate the disparity image I_D almost independently for each pixel by a WTA strategy applied to costs computed on local portions of the reference and target images, global stereo vision algorithms compute the whole disparity image I_D at once by imposing a smoothness model on scene depth distribution.

Such global algorithms generally adopt a Bayesian framework and model the disparity image as a Markov Random Field (MRF) in order to include within a unique framework cues coming from local comparisons between reference and target image and smoothness constraints. Global stereo vision algorithms typically estimate the disparity image by minimizing a cost function made by two terms:

$$\hat{I}_D = argmin_D[C_{data}(I_L, I_R, I_D) + C_{smooth}(I_D)] \tag{3}$$

$C_{data}(I_L, I_R, I_D)$ is the so-called "data term", representing the cost of a local matches (similar to the one of local algorithms). The sum of such costs over all the reference image points defines the cost of a disparity image $I_D \cdot C_{smooth}(I_D)$, called "smoothness term", defines the level of smoothness of disparity image I_D, by explicitly or implicitly accounting for discontinuities. $C_{smooth}(I_D)$ takes into account that scenes generally have quite flat disparity distributions except in presence of depth discontinuities, by penalizing disparity images that do not respect this type of behaviour. With a MRF model of the disparity image, $C_{smooth}(I_D)$ can be computed as sum of local terms accounting for the smoothness of neighbouring pixels. Other terms can be added to Eq. (3), in order to explicitly model occlusions and other a priori knowledge on the scene depth distribution.

Minimization (3) is not trivial, because of the great number of variables involved, i.e., $n_{row} * n_{col}$ disparity values of I_D, which can assume $d_{MAX} - d_{MIN} + 1$ possible values within range $[d_{min}, d_{MAX}]$. Therefore there are $(n_{rows} * n_{col})^{d_{MAX} - d_{MIN}+1}$ possible configurations of I_D. Since images acquired by current cameras can easily have millions of pixels within the range of hundreds of values, it is easy to understand how a greedy search for the minimum over all the possible configurations of I_D is not feasible. A classical solution to this is LBP, which searches for the minimum cost solution of (3) in a probabilistic sense. The disparity image is considered as a random field made by the juxtaposition of random variables (one for each pixel in I_D). Instead of optimizing the global probability density function defined on the whole random field, LBP marginalizes it, obtaining a probability density function for the disparity distribution of each point of I_D. The final optimization is performed by independently maximizing the marginalized probability density function at each point of I_D. The application of LBP to stereo vision has been proposed in [19]. An extensive description of LBP can be found in [20, 21]. An interesting perspective for the algorithms used for solving huge problems such as minimization (3) can be found in [20].

Global stereo vision algorithms are typically slower than local algorithms. However, by explicitly modelling the smoothness constraints (and by possibly including other constraints), they are able to cope with depth discontinuities and are more robust in texture-less regions.

1.5 Semi-Global Stereo Algorithm

Another very interesting class of stereo algorithms are the semi-global stereo approaches, which similarly to global methods adopt a global disparity model, but differently than global methods do not compute it on the whole disparity image in order to reduce the computational effort. More precisely the minimization of the cost function is computed on a reduced model for each point of I_D, differently than global approaches which estimate a whole disparity image I_D at once. For instance, the simplest semi-global methods, such as Dynamic Programming or Scan line Optimization [22] work in a 1D domain and optimize each horizontal image row by itself.

The so-called SGM algorithm [23] is a more refined semi-global stereo algorithm. It explicitly models the 3D structure of the scene by means of a point-wise matching cost and a smoothness term. Several 1D energy functions computed along different paths are independently and efficiently minimized and their costs summed up. For each point, the disparity corresponding to the minimum aggregated cost is selected. In [23] the authors propose to use 8 or 16 different independent paths. The SGM approach works well near depth discontinuities, however, due to its (multiple) 1D disparity optimization strategy, produces less accurate results than more complex 2D disparity optimization approaches. Despite its memory footprint, this method is very fast and potentially capable to deal with poorly textured regions.

2 Comparison of TOF Cameras and Stereo System

TOF cameras and stereo systems are both able to estimate the depth distribution of the acquired scene. Their estimates have different characteristics and clear-cut comparisons are not easy. Basic metrological quantities, such as accuracy, precision and resolution offer a systematic comparison ground considered in this section. The first part of the section recalls basic metrological definitions. The second part adapts them to the case of matricial depth measurements as produced by TOF cameras and stereo systems. Since actual comparisons between TOF cameras and stereo systems can only be made with respect to specific reference objects or scenes under the same illumination conditions, the last part of this section considers a practical example examined in this section in terms of well-known metrological quantities, i.e., accuracy, precision and measurement resolution. The performances of the stereo vision algorithms presented in the first section of this chapter (FW, LBP, SGM) and the metrological properties of the adopted sensors are analysed. Tests on real data are reported at the end of this section.

2.1 Fundamental Metrological Quantities

Let us first briefly recall the concepts of accuracy, precision and measurement resolution. For a more detailed presentation, the reader is referred to [24, 25].

Consider a measurement system S measuring a physical quantity Q. Assume the actual value of Q to be q^*. System S performs a series of n independent measurements of Q, all in the same experimental conditions. The values measured by S at each step are: $q_1, q_2, \ldots, q_N$.

Definition 1 *The accuracy* A of a measurement system S is the degree of closeness of measurements q_n to the actual value q^* of the quantity Q. It can be computed as the difference between the average on a set of measures of the same quantity and the actual value, i.e. $= |\bar{q} - q^*|$, where $\bar{q} = \frac{1}{N}\sum_{n=1}^{N} q_n$.

In the specific case of acquisition of depth maps, i.e., of depth information $z(p_{i,j})$ organized as an $I \times J$ matrix, as produced by TOF cameras and stereo vision systems, assume there are $z^n(p_{i,j})$, $n = 1, 2, \ldots, N$ depth map measurements of the scene Q available. In this case the accuracy of the measurement system is defined as

$$A = \frac{1}{I \cdot J}\sum_{i=1}^{I}\sum_{j=1}^{J}\left|\bar{z}(p_{i,j}) - z^*(p_{i,j})\right| \tag{4}$$

where $\bar{z}(p_{i,j}) = \frac{1}{N}\sum_{n=1}^{N} z(p_{i,j})$ and $z^*(p_{i,j})$ is the ground truth depth map.

Definition 2 *The precision* (or *repeatability*) P of a measurement system S is the degree to which repeated measurements under unchanged conditions show the same result. A common convention is to calculate the precision P of the system S in the measure of Q as the standard deviation of the measurement distribution σ_q of the measurements $q_1, q_2, \ldots, q_N$, i.e., $P = \sqrt{\frac{1}{N}\sum_{n=1}^{N}(q_n - \bar{q})^2}$, where $\bar{q} = \frac{1}{N}\sum_{n=1}^{N} q_n$.

The precision of a depth acquisition system can be computed by performing several depth measurements $z^n(p_{i,j}), n = 1, 2, \ldots, N$ and computing the standard deviation averaged over the whole depth map:

$$P = \frac{1}{I \cdot J}\sum_{i=1}^{I}\sum_{j=1}^{J}\sqrt{\frac{1}{N}\sum_{n=1}^{N}\left(z^n(p_{i,j}) - \bar{z}(p_{i,j})\right)^2} \tag{5}$$

where $\bar{z}(p_{i,j})$ is defined as above.

Definition 3 *The measurement resolution* R of a measurement system S is the smallest change δ_q in the underlying physical quantity with actual value q^* that produces a response in the measurement system.

2.2 Accuracy, Precision and Resolution of TOF Cameras and Stereo Systems

TOF cameras and stereo systems are rather different with respect to accuracy, precision and measurement resolutions as shown next.

2.2.1 Accuracy

With respect to accuracy, it is well known that TOF cameras depth measurements are characterized by a systematic offset caused by the harmonic distortion of the illuminators and camera pixels circuitry which generally varies with the distance and can be up to some tenths of centimetres (e.g., 400 mm, as reported in [26]). This means that in order to account for this artefact, one should provide an accuracy value for each distance value in the range of the measurable distances (e.g., in 500 – 5000 mm). However, for system characterization purposes, it is customary to synthetize the accuracy by a single value obtained by averaging the accuracy of the instrument over the range of measurable distances.

The TOF depth measurement offset due to harmonic distortion is of systematic nature and it can be reduced by a Look-Up-Table (LUT) correction independently applied to each pixel. However, since the measurement error depends also on the scene geometry and reflectance distribution, the LUT correction does not completely cancel the measurement error. The LUT-improved accuracy of a TOF camera is therefore limited. For example, according to the producer, the MESA Imaging SR4000 [27] is characterized by an accuracy of about 10 mm.

The accuracy of stereo vision systems depends both on the geometry of the setup and on the characteristics of the acquired scene (i.e., its geometry and the amount of texture information in the scene surfaces). The great variability of possible geometry and textures leads to non-systematic measurement errors which cannot benefit from simple strategy such as LUT-compensation. In order to better understand the origin of stereo systems error, let us consider the case of the FW stereo algorithm which, as shown in Fig. 3, for each reference image point tries to identify a conjugate point in a segment of the epipolar line in the target image. As already said, each couple of candidate conjugate points is characterized by a matching likelihood, quantified by a cost function (e.g., TAD). The more the two images are similar near to candidate conjugate points, the lower is their cost function and the more likely is the matching. The best case for stereo vision systems is when the scene characteristics are such that the local similarity between the *L* and the *R* images is high only in correspondence of the actual conjugate points pair (and low for the other candidate points pairs). In this case, the cost function has a minimum in correspondence of the conjugate points pair actually estimated by the WTA algorithm. This lucky situation requires that the reference and the target image satisfy the following two conditions:

- Reference and target image should exhibit an adequate amount of colour information (texture) near the actual conjugate points pair ("aperture problem")
- No other region of the target image along the epipolar line should be similar to the one corresponding to the actual target conjugate point ("repetitive texture pattern").

In case of insufficient texture or of multiple candidate conjugate points locally similar to the local reference image, there might be a disparity estimation mismatch with a consequent depth estimation error. Scene illumination greatly influences the possibility of this type of mismatches. Such depth measurement error does not grow regularly with the image noise, but tends to sudden bursts when the scene characteristics make the system unable to find the correct cost function minimum. The accuracy of the depth measurements produced by a stereo system is very hard to characterize by a single parameter since it strongly depends on the scene characteristic and on the used algorithms. All one can do is to define the accuracy of a stereo vision system for a specific scene or specific reference objects under specific illumination conditions (from different acquisitions of the same scene under the same conditions, and by computing the difference between the averaged estimated depth value at each pixel with respect to its actual depth value) as shown in Sect. 2.3. In general local stereo algorithms, totally dependent from the scene colour distribution, with respect to accuracy perform poorer than global and semi-global techniques less dependent on scene characteristics, because of the assumed smoothness model. At the same time it is clear that in case the actual scene does not match the assumed model, the assumptions behind global and semi-global methods turn against performance accuracy.

2.2.2 Precision

Since, as well known, the noise of TOF depth measurements can be assumed Gaussian [26], the depth measurement accuracy of TOF cameras relates directly to the mean of this Gaussian process, while the depth measurement precision is defined as its standard deviation. The standard deviation of the measurements is generally dependent on the actual depth and reflectance characteristics of the scene and on the background illumination at the IR wavelength at which the TOF camera operates. In particular, the standard deviation of the measurements increases as the distance from the object or the background illumination increase or the object reflectance decreases. For instance in the case of high reflectivity targets and low IR background illumination, the precision of the MESA Imaging SR4000 [27], according to the producer, is less than20 mm.

For the analysis of the precision of stereo systems let us consider the simple FW stereo algorithm. For simplicity denote by $D^l = Z^l(p_{i,j}), i = 1, 2, \ldots, I, j = 1, 2, \ldots, J$ the l-th depth map measurement. Assume, as shown in Fig. 4 that the scene is acquired Ntimes under same conditions giving the N images $I_L^1, I_L^2, \ldots, I_L^N$ from L, the N images $I_R^1, I_R^2, \ldots, I_R^T$ from R from which the N corresponding depth maps $D^1, D^2, \ldots, D^N$ are computed by the FW stereo vision algorithm.

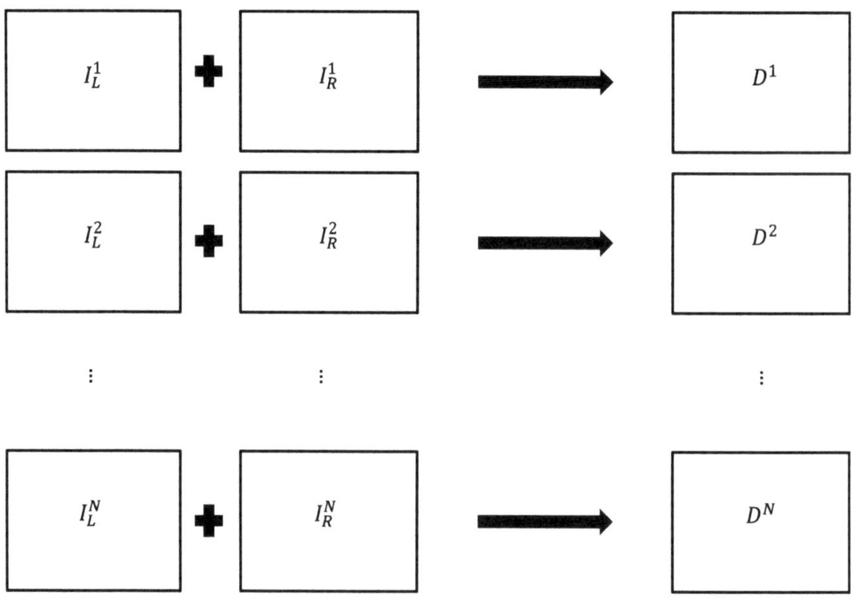

Fig. 4 Acquired stereo images and relative depth maps

The N depth maps are usually similar, but not identical due to the noise affecting images $I_L^1, I_L^2, \ldots, I_L^N$ and $I_R^1, I_R^2, \ldots, I_R^N$. Hence for a given point p_L the matching cost with respect to each candidate conjugate points varies for each acquisition. Noise fluctuations changing the conjugates pair that minimizes the matching cost change also the estimated depth map. The noise amount needed for changes of this nature clearly depends on the amount of texture in the scene. Low textured scenes are highly affected by image acquisition noise, while high textured scenes are less affected by it. The precision of FW stereo algorithm is directly related to scene reflectance characteristics and illumination conditions. Other stereo algorithms, such as SGM and BP are less noise-prone than FW, because the imposed scene model is generally capable to mitigate the noise influence.

The precision of a stereo vision system, with respect to a specific scene or reference object can be obtained from N acquisitions (as shown in Fig. 4) by computing the standard deviation of the measurements for each depth map point according to (5) as exemplified in Sect. 2.3.

2.2.3 Resolution

The measurement resolution of a matricial depth acquisition system, such as TOF cameras and stereo systems is characterized by spatial and depth resolution. The spatial resolution (or lateral resolution) for a fixed field-of-view (uniquely identified by the optics) is determined by the number image pixels and it represents the measurements resolution in the $x - y$ scene coordinates. The depth resolution, or resolution in the scene z coordinates, is the smallest scene variation δ_z capable to produce a depth response.

The spatial resolution of TOF cameras, i.e., the number of pixels in the sensor matrix, is currently considered one of their limitations, and it is one of the targets of TOF technology advancement. For instance, in the case of the MESA Imaging SR4000, the sensor matrix has 176×144 pixels. The analysis of a TOF camera depth resolution can be experimentally made as follows. Consider a set of N measurements of the TOF camera T positioned at a known distance z from a reference object, typically a plane of metrologically known characteristics. The minimum depth difference $\delta_Z(z)$ that produces a noticeable difference in the average of the depth measurements of two depth measures is the depth resolution of the camera T. Various factors may influence $\delta_Z(z)$, i.e., the sensitivity of the TOF cameras pixels, the precision of the sensor hardware and the final quantization grain of the depth measurements. Such a quantization grain is usually very fine. For example, the MESA Imaging SR4000 samples a depth interval of 5000 mm with 2^{14} values, i.e., with a quantization step of 0.3 mm. The other elements conditioning depth resolution cannot be treated analytically, and depth resolution must be estimated. As a practical example, the TOF resolution, for instance at $z = 1000$ mm, can be measured by taking a planar object and moving it from z to $z + \delta_z$ for smaller and smaller values of δ_z and by taking N measurement for each value of δ_z (e.g. with $N = 10^5$). If, for instance, at $\delta_z < 1$ mm the average of the TOF measurements at z and $z + \delta_z$ coincides and at $\delta_z^{'} = 1$ mm they do not coincide, it is possible to state that $\delta_z^{'} = 1$ mm is the resolution.

In the case of stereo systems the analysis of the measurements resolution can be done analytically. The spatial resolution of a stereo vision system is just given by the number of pixels of the left camera image sensor matrix. Since such matrices have a great number of pixels (e.g., 1032×778) stereo systems are considered high spatial resolution systems. This is certainly true, but it is also important to remind that stereo systems cannot estimate the depth value of all the points in their images and especially in presence of depth discontinuities they are not very precise. Furthermore it is possible to compute a disparity value only for samples visible by both cameras, e.g., usually the disparity cannot be estimated for the first columns on the side of the image or for points occluded with respect to one of the two cameras.

Concerning the depth resolution of stereo vision systems it is important to recall from (1) that the relationship between disparity and depth is not linear. Since the disparity is linearly sampled (the disparity for each pixel is an integer in the interval $[d_{MIN}, d_{MAX}]$), the relative depth values are non-linearly sampled.

Furthermore the quadratic dependence from depth values z of the depth increments Δz given by (2) has important consequences on depth resolution. Suppose that a point at depth z^* is acquired by a stereo system characterized by a focal f and a baseline b. The actual disparity value of that point is $d^* = \frac{bf}{z^*}$. The estimate $\hat{d}$ of d^* assumes only an integer value in $[d_{MIN}, d_{MAX}]$ which will be either $\lfloor \hat{d} \rfloor$ if $\hat{d} - \lfloor \hat{d} \rfloor \leq 0.5$, or otherwise $\lceil \hat{d} \rceil$ Consequently the estimate $\hat{z}$ of z might assume either value $\hat{z} = \frac{bf}{\lfloor d \rfloor}$ or $\hat{z} = \frac{bf}{\lceil d \rceil}$ and the minimum depth increment that the system can measure for a point at distance z^* is $\Delta z = \frac{bf}{\lfloor d \rfloor} - \frac{bf}{\lceil d \rceil}$. From (2) $\Delta z = \frac{z^{*2}}{fb} \Delta d$, where in this case $\Delta d = \lceil d \rceil - \lfloor d \rfloor = 1$. In other words the depth resolution decreases quadratically with the depth of the measured objects. Depth resolution can be improved by sub-pixel stereo matching, but the benefits are limited by classical interpolation artefacts. Sub-pixel techniques allow to reduce the value of Δd in (2) (e.g. $\Delta d \sim 0.1$), but cannot change the quadratic dependence of Δz with respect to depth z. Therefore TOF cameras usually have a better depth resolution Δz than stereo systems for distant objects and worse resolution than stereo systems for close objects.

Another important element to take into account is the computation time of the different scene depth estimation systems. While TOF cameras operation is simple and can be efficiently implemented in hardware, stereo algorithms, especially the global ones are computational complex. Rates of tents of depth estimates per second (e.g., 50 times per second) are typical of TOF cameras, while rates of few depth estimates per second are typical of software implementations of current stereo algorithms. Needless to say, the stereo vision algorithms speed can be greatly improved by hardware implementations.

2.3 Experimental Comparisons

In order to clarify the previous discussion some experimental comparisons of the performances of TOF cameras and stereo vision systems on a sample dataset are presented next. The reference scene showed by Fig. 5 has been acquired by both the TOF camera and the stereo acquisition system of the setup shown in Fig. 7. The scene depth map has then been estimated by three different stereo vision algorithms, namely, FW, SGM and LBP. The goal of this experiment is to give an example of how comparisons of this kind can be made in practice.

The implementations of the considered stereo vision algorithms can be found in the OpenCV library [9]. In particular, FW and SGM implementations are classical CPU stereo vision algorithms, while the considered LBP implementation exploits also GPU. A matching window of size 21×21 has been adopted for FW and SGM stereo vision algorithms, while a small 1×1 window for LBP. The scene was acquired $N = 10$ times by both the stereo system and the TOF camera. The three considered stereo vision algorithms have been applied to each stereo acquisition. In order to have a ground truth depth measurement the scene was also acquired by an active space-time stereo vision system [28, 29], with an accuracy of about 3 mm, way superior to that of both the stereo and the TOF camera. Fig. 6 shows three examples of estimated depth maps (one for each stereo algorithm) and the ground-truth depth-map computed by the space-time stereo.

Note that a depth measurement is not available for the pixels associated to 0 depth (black pixels) due to matching failure and occlusions in the case of the passive stereo algorithms and due only to occlusions in the case of space-time stereo.

The accuracy A and the precision P of the two systems were computed according to Eqs. (4) and (5) respectively using the space-time stereo data as ground truth and are shown in Table 1 together with the resolution characteristics.

Table 2 reports the execution times of the considered stereo algorithms:

It is worth reminding that the presented results apply to the considered reference scene and not to general scenes. Nevertheless they allow for some concrete

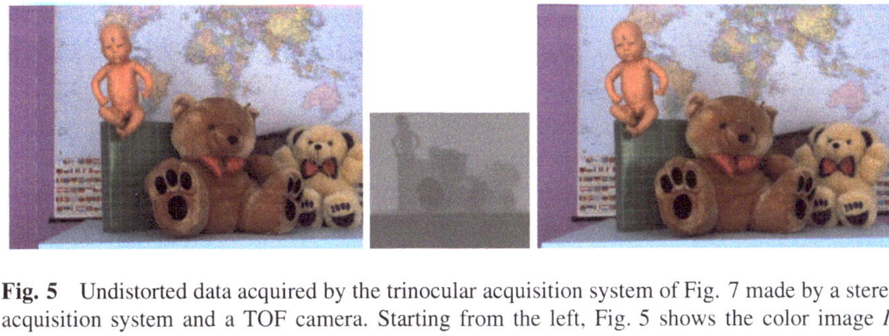

Fig. 5 Undistorted data acquired by the trinocular acquisition system of Fig. 7 made by a stereo acquisition system and a TOF camera. Starting from the left, Fig. 5 shows the color image I_L acquired by the left camera L of the stereo acquisition system, the depth map Z_T acquired by the TOF camera T and the color image I_R acquired by the right camera R of the stereo acquisition system

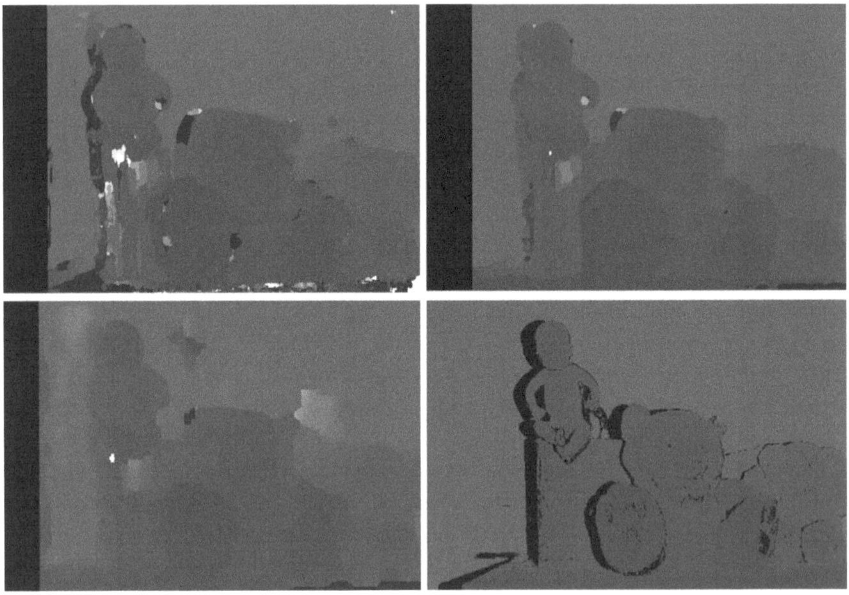

Fig. 6 Examples of depth maps estimated by FW (*up-left*), SGM (*up-right*) and LBP (*bottom-left*) stereo systems and ground truth depth-map acquired by space-time stereo (*bottom-right*)

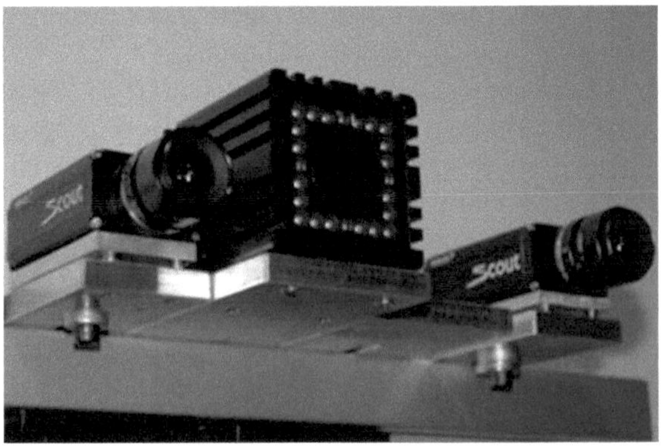

Fig. 7 An example of trinocular acquisition setup made by a TOF camera and a stereo vision system

and reasonable considerations of general kind based on quantitative data. Namely TOF cameras are typically faster and their results do not depend on the amount of texture information in the scene (with respect to the considered implementations). On the other side, stereo vision systems have better spatial resolution and allow more precise edge localization. Stereo depth resolution can be better than TOF resolution for closer objects. TOF cameras depth resolution is less dependent on the object distance than the one of stereo systems. In the case of stereo vision it is

Table 1 Experimental comparison between the TOF cameras and the stereo vision systems. Accuracy and precision are computed with respect to the scene shown in Fig. 5. Spatial resolution and depth resolution are characteristic of the considered acquisition system. The considered stereo system has focal $f = 856.3$ (pxl) and baseline $b = 176.8$ (mm)

	Stereo FW	Stereo SGM	Stereo LBP	TOF (MESA SR4000)
Accuracy	60 $[mm]$	35 $[mm]$	41 $[mm]$	30 $[mm]$
Precision	13 $[mm]$	2 $[mm]$	12 $[mm]$	2.6 $[mm]$
Spatial resolution	777×778	777×778	777×778	176×144
Depth resolution	$\frac{z^{*2}}{fb}$	$\frac{z^{*2}}{fb}$	$\frac{z^{*2}}{fb}$	$0.3[mm] < \delta_z < 1[mm]$

Table 2 Execution times of the stereo algorithms. FW and SGM are implemented on CPU, while LBP is implemented on GPU. The experiments were run on a machine with a 4 core Intel i7, $3.06[GHz]$ CPU and NVIDIA NVS 3100 M GPU

Stereo algorithm	Execution time [ms]
FW	~ 130
SGM	~ 2400
LBP	~ 1600

possible to change the baseline and focal in order to improve the resolution. The execution times of CPU and GPU stereo algorithms do not allow obtaining frame rates as high as those of TOF cameras.

Such complementary characteristics of the two systems open the way to the idea of fusing their data. A panoramic on the current state-of-the-art in TOF and stereo data fusion techniques is presented in the next section.

3 Fusion of Time-Of-Flight and Stereo Data

Section 2 shows that the characteristics of Time-Of-Flight cameras and of stereo systems are complementary in several aspects. The possibility of overcoming the limitations of the two kinds of acquisition systems by fusing their data has already received considerable attention. Since current stereo systems are relatively inexpensive with respect to the current TOF cameras this concept is rather attractive also for practical purposes.

Since the ultimate goal of fusing TOF and stereo information should be a system capable to provide depth information with high accuracy, precision and resolution, it is fair to say that the methods proposed so far can only fulfil a subset of such desired features.

The first part of this section considers the basic requirements of a set-up for jointly combining TOF and stereo data. Essential issues are geometrical configuration and mutual calibration between the two different acquisition systems. The rest of this section reviews the state-of-the-art fusion methods. The considered fusion algorithms are subdivided into local, global and semi-global.

3.1 Characteristics of the Acquisition Setup

The simplest acquisition setup for the joint exploitation of stereo vision and TOF information is a trinocular system made by a TOF camera and a pair of color cameras on a rig, as in the example of Fig. 7 made by a MESA Imaging SR4000 TOF camera together with two Basler video cameras. The TOF camera is normally between the two video cameras (usually closer to the reference camera) in order to reduce occlusions and to ensure the largest possible common region for the depth data acquired by the two systems. A schematic representation of the considered setup made by left camera L, right camera R and TOF sensor T in between them is shown in Fig. 8.

Left camera *L* and right camera *R* respectively act as reference and target camera also for the trinocular system of Fig. 8. The baseline of the stereo vision system and the focal of the stereo cameras are important factors for the final depth estimation quality. They can be selected upon the same considerations seen for stereo systems.

Synchronization, which can be made either by hardware or software, is very important for the performance of the trinocular system. Indeed one encounters

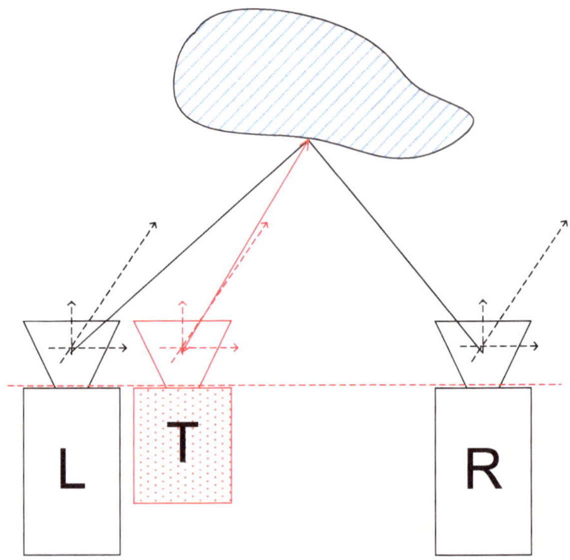

Fig. 8 Scheme of the trinocular acquisition setup made by a TOF camera and a stereo vision system

artefacts similar to the ones of stereo systems if the three cameras are not perfectly synchronized. The synchronization of the trinocular system is complicated by the fact that the acquisition frame-rate of L and R is generally different from that of T. The acquisition frame-rate of the trinocular system cannot be above the minimum between the acquisition frame-rate of the TOF and the acquisition frame-rate of the stereo cameras (without accounting for the execution time of the stereo algorithms). Either color or grayscale cameras can be used for L and R. Color cameras allow fusion algorithms that exploit color information. Color fusion algorithms are generally more precise and accurate but more resource-consuming than gray-level fusion algorithms, as already seen for stereo vision algorithms.

For information fusion purposes the TOF and the stereo system must be jointly calibrated. Each of the three cameras L, R and T is associated to its own 3D reference system as shown in Fig. 6. Each camera has specific projection properties and TOF cameras can be assumed similar to standard cameras in this respect. The systematic error of the TOF depth measurements must be taken into account for the calibration step. The calibration of the trinocular system requires to estimate the intrinsic parameters of L, R and T, the relative positions and orientations (rototranslations) of their reference systems (or extrinsic parameters) and the systematic error of the depth measurements of T. The intrinsic and extrinsic parameters of cameras L and R, can be obtained by classical stereo calibration tools. Standard camera calibration methods are not effective for estimating the intrinsic parameters of TOF cameras due to issues like their limited spatial resolution, the poor quality of the acquired images (just grayscale reflectance maps at the TOF wavelength), the high noise levels and the severe lens distortion. How to properly calibrate the intrinsic parameters of a TOF camera can be found in Chap. 5.

The relative roto translations between the colour cameras and the TOF camera reference system in principle can be estimated by standard stereo calibration techniques applied to the three possible couples of cameras. In practice this does not lead to accurate calibrations because of the differences in the pairs made by a standard camera and a TOF camera (image resolution, distortion, image noise). Better results can be obtained by jointly registering the three devices. The accuracy of this global optimization approach is limited by the localization accuracy of the TOF camera. The best results are obtained by exploiting the 3D information from the TOF camera. The approaches of [30] and [31] perform first a standard stereo calibration for the stereo system. The calibrated stereo system is then used to acquire the 3D location of the corners of a target checkerboard. The TOF camera is also used to acquire the 3D location of the checkerboard corners. The two point clouds corresponding to the checkerboard corners, one coming from the stereo system and the other from the TOF camera, expressed with respect to the two reference systems are then registered together by Horn algorithm [32] within a RANSAC [33] framework. This procedure gives good estimates of the roto translation between the two reference systems. Mirroring the structure adopted in the presentation of stereo vision algorithms, the rest of this section first presents local fusion algorithms, then global fusion algorithms, and finally semi-global fusion algorithms.

3.2 Local Approaches

A first simple way of fusing TOF and stereo data, presented in [34] is to just average the depth measures of the two subsystems. In this approach, the depth map acquired by the TOF and the one acquired by the stereo pair are separately obtained, then registered on a unique reference system and finally averaged. The quality of the final results obtained by this method is rather limited, since the errors of the two acquisition systems add up. Furthermore, averaging does not take into account the different reliability of TOF data and of stereo reconstruction in different scene context (e.g., stereo reconstruction is more reliable for textured than for non-textured surfaces, etc.). The final spatial resolution increases up to the resolution of the stereo vision system only fictitiously, since the TOF depth-map is simply interpolated, and interpolation artefacts propagate to the final depth map through the averaging operation.

In the more interesting method proposed by [35] the depth map acquired by the TOF is firstly upsampled by a hierarchical application of bilateral filtering. A plane-sweeping stereo algorithm is then applied to the acquisition volume defined with respect to the TOF reference system. Finally the depth measures acquired from both the TOF camera and the stereo algorithm are fused together by means of a confidence based strategy. This approach is quite interesting and suited for GPU implementation, but the simple plane-sweeping stereo algorithm does not improve much upon the results obtained just by interpolating the TOF data by bilateral filtering. Such a method provides a scene depth estimate characterized by high spatial resolution. Accuracy and precision are also improved (up to extents, comparable with what is at reach of global and semi-global methods).

As previously said the precision and accuracy of the data coming from stereo systems and from the TOF cameras depend on many different factors and an effective fusion approach should take into account the different reliability of the data of the two systems in different scene contexts. Some local [30] and global methods [31] exploit a probabilistic framework in order to achieve this target. The idea is to build for each sample in the scene two probability functions representing the likelihood of a given depth measure Z given by the TOF camera and by the stereo system. Let us denote with $P[Z|M_T]$ the probability of having a certain depth value Zgiven the measureM_T by the TOF camera and with $P[Z|M_S]$ the probability of Z given the measure M_S by the stereo system. The fusion problem can be expressed in a maximum a posteriori (MAP) formulation as the search for the depth value Zthat maximizes the posterior probability given the TOF and the stereo measures M_T and M_S, i.e.,:

$$\hat{Z} = argmax_Z(P[Z|M_T, M_S]) \tag{6}$$

In most practical situation it is possible to assume independent the TOF and stereo measures [27] and to compute the estimated depth as the value that maximizes the product of the TOF and stereo system probabilities, i.e.:

$$\hat{Z} = argmax_Z(P[Z|M_T]P[Z|M_S]) \tag{7}$$

As expected the critical issue for this kind of approaches is the construction of the two probability functions $P[Z|M_T]$ and $P[Z|M_S]$.

The TOF probability function depends on the precision of the TOF measures that, as already discussed in this book, is a function of the different sources of noise that affects the TOF camera and of the scene geometry and reflectance characteristics. As pointed out in [26] the TOF noise can be approximated by a Gaussian distribution. In order to estimate the standard deviation of the distribution it is necessary to take into account the properties of the employed TOF camera but also the fact that the signal-to-noise ratio depends on the strength of the received signal which is a function of the object reflectance. In [34] this issue is solved by a statistical analysis of the TOF measures reliability with different reflectances that is then used to build a function that represents the standard deviation of the noise as a function of the reflectance. Some TOF cameras like the MESA Imaging SR4000 also return a confidence map (that in fact mostly depends on the strength of the returned signal) associated to the measurements reliability, used for example in [30] to compute the standard deviation of the Gaussian probability distribution of the TOF noise. It is also important to remind that TOF measures are usually less accurate near scene depth discontinuities because of limited spatial resolution and scattering. In order to take this issue into account one may take into account the local variance in the computation of the probability function.

The stereo probability model depends on the employed stereo algorithm. Since most stereo algorithms associate a matching cost to the disparity of each candidate (e.g., TAD) the probability of each depth value can be computed as a function of the matching cost. Note how this approach implicitly takes into account the reliability of the stereo algorithm, e.g., in uniform texture-less areas the matching cost and the probability function are quite low forcing the use of TOF data in these regions while in highly textured areas it is just the opposite.

Once the two probability functions are available, for each sample it is possible either to just select the depth value that maximizes their product or to include them into any global probabilistic approach, as it will be shown in the next sub-section. The fusion algorithm proposed in [30] aims at improving the depth accuracy and resolution of both the stereo and the TOF depth measurements. Since it intrinsically accounts for the confidence of the measurements produced by the two subsystems, it also improves accuracy and precision (Fig. 9 shows an example of the results of this approach). It does not improve though the spatial resolution of the estimated depth map which remains the one of the TOF.

(a) (b) (c) (d)

Fig. 9 Fusion of stereo and TOF data exploiting the local approach of [29]: **a** picture of the framed scene; **b** depth map acquired by the TOF camera; **c** depth map after the fusion of the two data sources; **d** difference between the two depth maps

3.3 Global Approaches

The method proposed in [31] and its extension proposed in [36] are based on a global MRF formulation, in which a belief propagation algorithm optimizes a global energy function. This method aims at increasing both resolution and accuracy of the depth measurements performed by each single subsystem. Unfortunately the global optimization procedure makes it rather slow. It does not clearly improve upon depth resolution.

The approach previously introduced can easily incorporate probabilistic frameworks such as MRF models usually employed in global stereo vision algorithms. Global stereo vision algorithms usually rely on a MRF model where the employed probability functions model the likelihoods given by the different clues. With respect to the Bayesian model of Eq. (3), the basic idea in order to include the TOF measures is to add a further data probability cost function that models the TOF probability. In this way, together with the smoothness term, there are two data terms, one depending on the stereo matching cost and one on the TOF data. The final cost problem can be expressed as:

$$\hat{D} = argmin_D[C_T(M_T) + C_S(M_S) + C_{smooth}(D)] \tag{8}$$

where $\hat{D}$ is the depth map produced by the fusion algorithm, C_T the cost of the TOF camera T, function of the measures M_T of T, C_S the cost of the stereo system S, that depends on the cost function M_S of the adopted stereo algorithm S, and C_{smooth} the smoothness term cost, that depends on the actual scene depth-map D. Problem (8) can be solved by the same techniques used in stereo vision methods such as Graph-Cuts or LBP.

A temporal extension of this method, proposed in [37], forces the consistency between current, previous and subsequent frames by computing the matching cost of the corresponding pixels in the three frames after optical flow estimation. The optimization in this case is solved by LBP (Fig. 10).

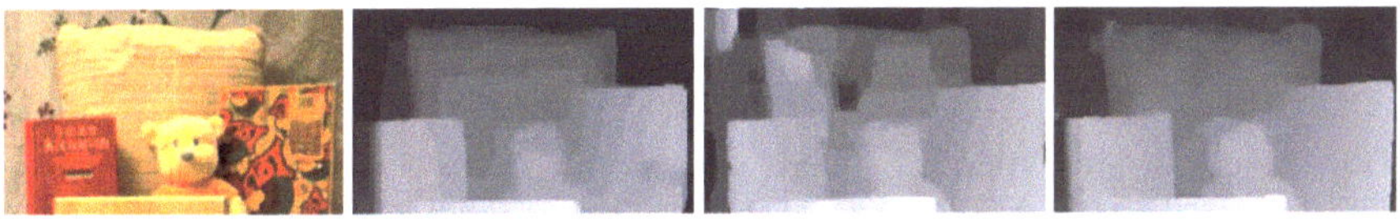

Fig. 10 Fusion of stereo and TOF data exploiting the global approach of Zhu et Al. [36] *(courtesy of the authors)*: **a** picture of the framed scene; **b** depth map acquired by the TOF camera; **c** depth map acquired by the stereo system; **d** depth map after the fusion of the two data sources

3.4 Semi-Global Approaches

A first simple semi-global approach is the one by [38], in which the depth map acquired by the TOF is firstly reprojected on the reference image of the stereo pair, it is then interpolated and finally used as initialization for the application of a dynamic programming stereo algorithm. The main limit of this method is that it enforces the so-called ordering constraint that is not always satisfied. A violation of such a constraint usually leads to severe artefacts. Such a method therefore produces high spatial resolution, but it does not considerably improve accuracy and precision.

Similarly to [30] also the approach of [39] builds a cost function for each sample that depends both on the TOF measures and on the stereo matching cost (computed in this case by the method of [40]). The algorithm firstly reprojects the depth measures coming from the TOF on the two cameras images and then computes the stereo matching cost. If the matching cost is below a pre-defined threshold the TOF measure is considered valid and the cost function for that sample is approximated by a reverse Gaussian distribution centered at the TOF measurement. Otherwise the TOF measure is considered wrong and the cost is computed only from the matching cost of the stereo algorithm. Differently from [30] there is also a final cost aggregation stage based on the semi-local method of [23] that employs a smoothness constraint and aggregates the costs over 16 surrounding 1D directions as described in Sect. 1.5. This approach propagates the depth measures of the TOF on the high resolution lattice of the stereo cameras and is able to produce high resolution depth maps. It is also quite robust but differently from other schemes it basically uses stereo data only to replace the TOF measures when not available or not reliable instead of really combining the two measures. This makes this method suitable for situations where the TOF data are more reliable than the stereo ones. When the two systems have similar performances it is probably not the best solution.

4 Conclusions

TOF cameras are emerging depth measurement instruments. They are very attractive because of their speed and their robust behaviour in terms of dependence from scene characteristics (e.g., texture distribution). Stereo systems are classical depth measurement instruments, very attractive for the inexpensiveness of their hardware components and the large body of stereo algorithms available in the literature. The variety of stereo algorithms gives great flexibility to stereo systems.

The comparison of these conceptually and technologically different depth measurement methods is the first issue considered by this chapter. It is first approached on the basis of classical metrological concepts such as accuracy, precision and resolution. However because of the great number of involved system, environment and scene factors, the comparison can only be experimentally made as exemplified at the end of Sect. 2.

The data obtained by the two depth measurement systems, as shown in Sect. 2, have characteristics making worthwhile considering their fusion. Section 3 reviews state-of-the-art techniques for combining data coming from both a TOF camera and a stereo system. In this connection stereo vision algorithms can be revisited and extended in order to profitably take into account the characteristics of TOF data. As seen from Sect. 3, the investigation of the fusion between TOF camera and stereo data has already received considerable attention, but there is still plenty of room for improvements.

It is worth noting that the comparison between TOF camera and stereo data and their fusion are deeply related topics since a data fusion approach is profitable only if it concerns data with substantial differences and possibly somehow complementary characteristics.

References

1. T. Luhmann, S. Robson, S. Kyle, I. Harley, *Close Range Photogrammetry: Principles, Techniques and Applications*, (Wiley, Chichester, 2007), pp. 528 ISBN 978-0-47010-633-4
2. E.M. Mikhail, J.S. Bethel, J.C. McGlone, *Introduction to Modern Photogrammetry*, (Wiley, New York, 2001), pp. 496 ISBN 978-0-47130-924-6
3. Point Grey Research, Inc., http://www.ptgrey.com/products/stereo.asp
4. TYZX, Inc., http://www.tyzx.com
5. Middlebury Stereo Vision Page, http://vision.middlebury.edu/stereo/
6. R. Szeliski, *Computer Vision, Algorithms and Applications, Series: Texts in Computer Science*, 1st edn. (Springer, New York, 2011) pp. 812 ISBN 978-1-84882-934-3
7. G. Bradski, A. Kaehler, *Learning OpenCV: Computer Vision with the OpenCV Library*, 1st edn. (O'Reilly Media, Sebastopol, 2008) pp. 555 ISBN 978-0596516130
8. J.-Y. Bouguet, Camera Calibration Toolbox for Matlab, http://www.vision.caltech.edu/bouguetj/calib_doc/
9. OpenCV, Open Source Computer Vision library, http://opencv.willowgarage.com/wiki/

10. A. Fusiello, E. Trucco, A. Verri, A compact algorithm for rectification of stereo pairs, machine vision and applications 12(1) (Springer Berlin/Heidelberg, 2000) pp. 16–22 ISSN: 0932-8092
11. D. Scharstein, R. Szeliski, A taxonomy and evaluation of dense two-frame stereo correspondence algorithms. Int. J. Comput. Vis. 47(1–3) (Kluwer Academic Publishers, Hingham, 2002), pp. 7–42 ISSN 0920-5691
12. A.W. Gruen, Adaptive least squares correlation: a powerful image matching technique. South African J Photogramm, Remote Sens. Cartogr. **14**, 175–187 (1985)
13. F.C. Crow, Summed-area tables for texture mapping, Proceedings of the 11th annual conference on Computer graphics and interactive techniques, SIGGRAPH 84, (ACM, New York, 1984), pp. 207–212 ISBN 0-89791-138-5
14. M.J. McDonnell, Box-filtering techniques, computer graphics and image processing, 17(1), 65–70 (1981). ISSN 0146-664X, 10.1016/S0146-664X(81)80009-3
15. T. Kanade, M. Okutomi, A stereo matching algorithm with an adaptive window: theory and experiment, IEEE transactions on pattern analysis and machine intelligence, **16**, N. 1 920–932, (1994)
16. A. Fusiello, V. Roberto, E. Trucco, Symmetric stereo with multiple windowing. Int. J. Pattern Recognit. Artif. Intell. **14**, 1053–1066 (2000)
17. K.-J. Yoon, I.S. Kweon, Adaptive support-weight approach for correspondence search. IEEE Trans. Pattern Anal. Mach. Intell. **28**(4), 650–656 (2006)
18. F. Tombari, S. Mattoccia, L. Di Stefano, Segmentation-based adaptive support for accurate stereo correspondence, in *Proceedings of IEEE Pacific-Rim Symposium on Image and Video Technology*, vol 1 December 17–19 (Santiago, Chile, 2007), pp. 427–438, Springer
19. J. Sun, N.-N. Zheng, H.-Y. Shum, Stereo matching using belief propagation. IEEE Trans. Pattern Anal. Mach. Intell. **25**(7), 787–800 (2003)
20. R. Szeliski, R. Zabih, D. Scharstein, O. Veksler, A. Agarwala, C. Rother, A comparative study of energy minimization methods for Markov random fields, ECCV, 2006, pp. 16–29
21. S.Z. Li, *Markov random field modelling in image analysis (Advances in Pattern Recognition)*, 3rd edn. (Springer, London, 2009). ISBN 1848002785
22. I.J. Cox, S.L. Hingorani, S.B. Rao, B.M. Maggs, A maximum likelihood stereo algorithm. Comput. Vis. Image Underst. **63**, 542–567 (1996)
23. H. Hirschmuller, Stereo processing by semiglobal matching and mutual information. IEEE Trans. Pattern Anal. Mach. Intell. **30**(2), 328–341 (2008)
24. M. Burns, G.W. Roberts, *Mixed-Signal IC test and measurement (The Oxford series in electrical and computer engineering)* (Oxford University Press, New York, 2001)
25. Evaluation of measurement data—Guide to the expression of uncertainty in measurement, Joint Committee for Guides in Metrology, JCGM 100:2008
26. T. Kahlmann, H. Ingensand, Calibration and development for increased accuracy of 3D range imaging cameras, J. Appl. Geodesy, **2**(1), 1–11 (2008), ISSN (Online) 1862-9024, ISSN (Print) 1862-9016
27. MESA Imaging, http://www.mesa-imaging.ch/
28. J. Davis, D. Nehab, R. Ramamoorthi, S. Rusinkiewicz, Spacetime stereo: A unifying framework for depth from triangulation, CVPR, 359–366 (2003)
29. L. Zhang, B. Curless, S. M. Seitz, *Spacetime stereo: Shape recovery for dynamic scenes*. In IEEE Computer Society, CVPR, 2003, pp. 367–374
30. C. Dal Mutto, P. Zanuttigh, G.M. Cortelazzo, *A Probabilistic Approach to TOF and stereo data fusion, 3DPVT* (France, Paris, 2010)
31. J. Zhu, L. Wang, R. Yang, J. Davis, Fusion of Time-Of-Flight depth and stereo for high accuracy depth maps. CVPR, 23–28 June 2008
32. B.K.P. Horn, Closed-form solution of absolute orientation using unit quaternions. J. Opt. Soc. America **4**(4), 629–642 (1987)
33. M.A. Fischler, R.C. Bolles, Random Sample Consensus: A paradigm for model fitting with applications to image analysis and automated cartography. Comm. ACM **24**, 381–395 (1981)

34. K.-D. Kuhnert, M. Stommel, Fusion of stereo-camera and PMD-camera data for real-time suited precise 3D environment reconstruction, IEEE/RSJ international conference on intelligent robots and systems, 9–15 Oct 2006, pp. 4780–4785
35. Q. Yang, K.-H. Tan, B. Culbertson, J. Apostolopoulos, Fusion of active and passive sensors for fast 3D capture, IEEE international workshop on multimedia signal processing (MMSP), pp. 69–74, 4–6 Oct 2010
36. J. Zhu, L. Wang, R. Yang, J. E Davis, Z. Pan, Reliability fusion of Time-Of-Flight depth and stereo geometry for high quality depth maps, IEEE Trans. Pattern Anal. Mach. Intell. 33(7), ISSN 0162-8828, (IEEE Computer Society, Washington, 2011) pp. 1400–1414
37. J. Zhu, L. Wang, J. Gao, R. Yang, Spatial-temporal fusion for high accuracy depth maps using dynamic MRFs. IEEE Trans. Pattern Anal. Mach. Intell. **32**(5), 899–909 (2010)
38. S. A. Gudmundsson, H. Aanaes, R. Larsen, Fusion of Stereo Vision and Time-Of-Flight Imaging for Improved 3D Estimation. Int. J. Intell. Syst. Technol. Appl. 5(3/4), 425–433, (2008) ISSN 1740–8865, (Inderscience Publishers, Geneva, Switzerland 2008)
39. J. Fischer, G. Arbeiter, A. Verl, Combination of Time-Of-Flight depth and stereo using semiglobal optimization, IEEE international conference on robotics and automation (ICRA), pp. 3548–3553, 9–13 May 2011
40. S. Birchfield, C. Tomasi, A pixel dissimilarity measure that is insensitive to image sampling, IEEE Trans. Pattern Anal. Mach. Intell. **20**(4), 401–406, N. 6 (1998), ISSN 0162-8828
41. C.M. Bishop, *Pattern recognition and machine learning (Information Science and Statistics)* (Springer, New York, 2006). ISBN 0387310738, 9780387310732

TOF Cameras in Ambient Assisted Living Applications

Alessandro Leone and Giovanni Diraco

1 Introduction

In the recent years, the phenomenon of population ageing is receiving increasing attention firstly for healthcare and social impacts (rising health-care costs, life-style changes, etc.) and secondly as an opportunity to leverage the full potential of technology in making automated services for lonely elderly people. In this vision, AAL has been introduced as a term describing solutions based on advanced ICT technologies to support conduct of life. Relevant applications in this field relate, for instance, to the prevention and detection of potential dangerous events such as falls in the elderly, integrated in a wider emergency system which may help in saving lives. On the other hand, many AAL applications, especially in the homecare field, exploit the inference of human activities in order to support the everyday living of elderly people. Applications herein are devoted to support a wide range of needs from specific rehabilitation exercises to better insights into how perform the so called Activities of Daily Living (ADLs), helping geriatricians to evaluate the autonomy level of older adults by employing a variety of electronic aids and sensors [1]. The design of such AAL applications is normally based on paradigms of ambient intelligence and context-awareness providing intelligent environments in which various kind of sensors are deployed. Typical adopted sensors are accelerometers, gyroscopes, video cameras, microphones, pressure switches, and so on. Solutions based on these sensors can roughly be grouped on the basis of their operation modality into three main categories: ambient-based, wearable-based and camera-based solutions [2, 3]. The ambient category relates to those sensors that are embedded into appliances or furniture in order to detect

A. Leone (✉) · G. Diraco
Consiglio Nazionale delle Ricerche—Istituto per la Microelettronica ed i Microsistemi, Via Monteroni, presso Campus Universitario, Palazzina A3, Lecce, Italy
e-mail: alessandro.leone@le.imm.cnr.it

G. Diraco
e-mail: giovanni.diraco@le.imm.cnr.it

F. Remondino and D. Stoppa (eds.), *TOF Range-Imaging Cameras*,
DOI: 10.1007/978-3-642-27523-4_10,

presence, door open/close, etc. They require typically an ad hoc design or redesign of the home environment. Wearable devices are based essentially on accelerometer and/or gyroscope sensors. This solution does not require any environmental modification since devices are worn by the user; however wearable devices are prone to be forgotten or worn in a wrong body position, exhibiting a low acceptance rate. Camera-based solutions require the installation of at least one camera in each monitored room allowing the capture of the most of the activities performed and avoiding, at the same time, a large number of ambient-based sensors. Furthermore, apart from being non-invasive camera provides a rich and unique set of information that cannot be obtained from other types of sensors.

Aiming to highlight the benefits of TOF-based RIM in relevant AAL contexts, this chapter focuses on two central AAL scenarios, namely the critical event detection and the analysis of human activities. In particular, the fall detection is considered as the main representative application within the first scenario, whereas the problem of posture recognition is faced within the second one since it is a fundamental prerequisite to all kind of human activity inferences. The main principles of the different approaches are discussed with less of a focus on theoretical details, instead methodologies and their integration into practical implementations are suggested, giving realistic hints on how to handle the main technical issues typical of AAL contexts. The presented methodologies are implemented by using a state-of-the-art TOF range camera and a very compact embedded PC. The design of the suggested system takes into account the ethical aspects in order to maximize the user's acceptance rate and minimize the risk of loss of privacy.

The chapter is organized as follows. In Sect. 2, the active vision is compared with the passive vision and the advantages of the first in AAL contexts are highlighted. A full automated system for detection of falls in the elderly by using TOF vision is presented in Sect. 3, in which both methodological and technical issues are considered. In Sect. 4, the presented framework is extended suggesting a TOF-based solution for human posture recognition well suited for AAL contexts. Finally, the Sect. 5 concludes the chapter by discussing the proposed framework and giving some final considerations.

2 Advantages of Active Vision in AAL Contexts

Generally, the usage of monocular vision for surveillance and monitoring purpose is considerably troublesome since a single camera view can be strongly affected by perspective ambiguity when the viewpoint is unfavorable [4, 5]. The stereo vision (or in general the multiple view vision in which two cameras or more capture the same scene) overcomes perspective problems exploiting 3D geometric representation of human shape. However, stereo/multiple view vision deals with the ill-posed problem of the stereo correspondences strongly affected by poor textured regions and violation of brightness constancy assumption. In addition, the usage of multiple cameras requires both intrinsic and extrinsic camera calibrations that unfortunately are time consuming and error prone activities [6, 7]. Moreover, both monocular and stereo/multiple view vision systems fall within the so called passive vision in which the vision system measures the visible radiation already present in the scene due to natural or artificial illuminations. In general passive vision is well-known to be demoted by many factors such as the presence of shadows, camouflage effects (overlapped regions having similar colors), brightness fluctuations, few surface cues (poor textured regions) and occlusion handling. Recently, the active vision, mainly by using TOF cameras, is increasingly investigated in order to overcome the drawbacks of passive vision systems [8–16]. The manufacture costs of active vision systems in general and TOF cameras in particular are decreasing thanks to a lot of researches in progress especially gained by gaming industry strongly interested in new Natural User Interface: in a near future these devices are likely to be as cheap as webcams are today [11, 17]. Table 1 synthesizes the most important characteristics of TOF sensors in comparison with passive stereo vision systems. The main advantage in the use of TOF is the description of a scene with a more detailed information, since both depth map and intensity image can be used at the same time. In particular, previously mentioned problems of passive vision (foreground camouflage, shadows, partial occlusions, etc.) can be overcome by using depth information that is not affected by illumination conditions and objects appearance. Although the passive stereo vision provides depth information in a less expensive way, this approach presents high computational costs and it fails when the scene is poorly textured or the illumination is insufficient; vice versa, active vision provides depth maps even if appearance information is poor textured and in all illumination conditions [18, 19]. However, it is important to note that both distance and amplitude images delivered by the TOF camera have a number of systematic drawbacks that must be compensated. The main amplitude-related problem comes from the fact that the power of a wave decreases with the square of the distance it covers. For the previous consideration, the light reflected by imaged objects rapidly decreases with the distance between object and camera. In other words, objects with the same reflectance located at different distances from the camera will appear with different amplitudes in the image. Furthermore, in several situations active vision may exhibit unwanted behaviors due to limitations of specific 3D sensing technology (limited depth range due to aliasing, multi path, reflection object properties) [20]. Benefits of TOF sensors in surveillance contexts are summarized in Table 2, whereas drawbacks are reported in Table 3.

In order to understand the advantages in the use of range imaging in surveillance, a qualitative comparison between intensity-based and depth-based segmentation is presented in Fig. 1, using the same well-known segmentation approach (Mixture of Gaussians-based background modelling and Bayesian framework for segmentation) [21]. The two images, intensity and range respectively, are taken by the same TOF camera at the resolution of 176 × 144 pixels. The better segmentation is achieved by using the depth image, whereas the same segmentation approach applied on the intensity image suffers of mimetic effects.

Table 1 Comparison of important characteristics of TOF cameras and stereo vision systems

	TOF sensor	Stereo (passive) vision
Depth resolution	Sub-centimetre (if chromaticity conditions are satisfied)	Sub-millimetre (if images are highly textured)
Spatial resolution	Medium (QCIF, CIF)	High (over 4CIF)
Portability	Dimensions are the same of a normal camera	Two video cameras are needed and also external light source
Computational efforts	On-board FPGA for phase and intensity measurement	High workload (the calibration step and the correspondences search process are hard)
Cost	High for a customizable prototype (1000–3000€)	It depends on the quality of stereo vision system

Table 2 Advantages in the use of TOF sensors in surveillance contexts

	TOF sensor	Passive vision
Illumination conditions	Accurate depth measurement in all illumination conditions	Sensible to illumination variations and artificial lights. Unable to operate in dark environments
Shadows presence	It does not affect principal steps of monitoring applications	Reduced performances in segmentation, recognition, etc
Objects appearance	Camouflage is avoided but appearance could affect depth precision (chromaticity dependence)	Camouflage effects are presented when foreground/background present same appearance properties
Extrinsic calibration	Not needed when only one camera is used	Always needed

Table 3 Drawbacks in the use of TOF sensors in surveillance contexts

	Drawback description
Aliasing	It affects the non-ambiguity range i.e., the maximum achieved depth is reduced (up to 7.5 m)
Multi-path effects	Depth measurement is strongly corrupted when the target surface presents corners
Objects reflection properties	Materials having different colors exhibit dissimilar reflection properties that affect reflected light intensity and, therefore, depth resolution
Field of view	Usually it is limited so that an accurate positioning of the sensor is needed. A pan-tilt architecture could be useful

Moreover, the use of only depth images for measuring allows to improve the preprocessing process and, at the same time, to guarantee the person's privacy since chromatic information is not acquired: only depth measurements are sufficient to detect body movements and postures.

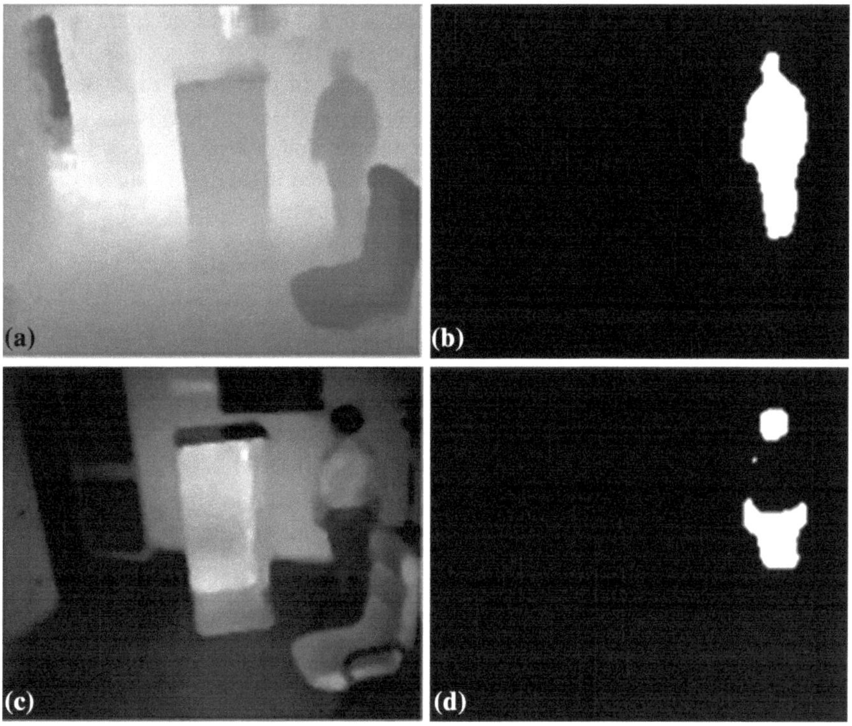

Fig. 1 Segmentation results are shown (**b**, **d**) when the segmentation approach in [39] is applied on depth (**a**) and intensity (**c**) information, respectively. The better segmentation is achieved starting from the depth image, whereas the same segmentation approach applied on the intensity image suffers of mimetic effects (sweater and wall present the same brightness)

3 A TOF Camera-Based Framework for Fall Detection

Actually, the problem of falls in the elderly has become a healthcare priority in all industrialized countries around the world due to the related high social and economic costs [22]. The consequences of falls in elderly may lead to psychological trauma [23], physical injuries [24], hospitalization and death in the worst case [25]. The medical importance of automatic fall-detection is apparent if the two following aspects are taken into account: (a) the involuntarily remaining on the floor for a long period after a fall is related with the morbidity/mortality rate [26]; (b) the elderly may not be able to activate a Personal Emergency Response Systems (PERS) due to the potential loss of consciousness [27]. The most investigated camera-based approach is the monocular one in which a camera alone captures the image frames. A monocular approach was investigated by Shaou-Gang et al. [28], detecting falls by measuring the aspect ratio of the bounding box of the body. Instead, Jansen and Deklerck [29] used depth maps obtained by a stereo camera system to detect inactivity by estimating the orientation change of the body.

A manually calibrated multiple camera approach was used by Cucchiara et al. [30] in order to detect a fall by inspection of the 3D body shape. Since passive vision (both monocular and stereo based) systems suffer of previously discussed drawbacks, recently authors start to investigate the problem of detection of falls by using active vision [12, 16] also in conjunction with other kind of sensors [14]. The suggested methodology for fall detection is discussed in the following subsections starting with the description of the hardware platform.

3.1 The Hardware Platform

The hardware platform used in the fall-detection framework includes two main components: an embedded PC equipped with an Intel® Atom™ Processor and managed by a Linux-based OS, and the MESA SR3000 [31] TOF camera installed in a wall mounting static setup as discussed in the following subsection. The extrinsic camera calibration is performed in a fully automated way by using a self-calibration procedure (see Sect. 3.2) in order to meet the easy-to-install requirement, whereas the intrinsic calibration is not required since the camera comes intrinsically calibrated by manufacturer.

3.2 Camera Mounting Setup

In this subsection the mounting setup of a TOF camera is discussed. The best camera mounting setup can be defined taking into account the following constraints: (1) the camera is static to limit the computational cost of a pan/tilt handling algorithm; (2) a people height of 1.75 ± 0.20 m is assumed. The two camera mounting configurations investigated were both ceiling and wall mounting setup. The Fields-of-View FoVw (for wall mounting) and FoVc (for ceiling mounting) can be quantitatively compared assuming that the following quantities are given: the covered room length L, the room height H, the average people height h. The two planes ρ and π are considered in order to evaluate the effective camera Field-of-View (Fig. 2): the first plane is referred to the ground floor, whereas the second one is referred to the people head position. In particular FoVw and FoVc are constrained in order to capture the whole π plane. Defined the ratio between the room length L and the distance H–h from the people head position to the ceiling, that is $L' = L/(H\text{–}h)$, the FoVw and FoVc are computed by using the following relations:

$$FoV_c(L') = 2\tan^{-1}\left(\frac{L'}{2}\right), \qquad FoV_w(L') = \frac{\pi}{2} - \tan^{-1}\left(\frac{1}{L'}\right). \quad (1)$$

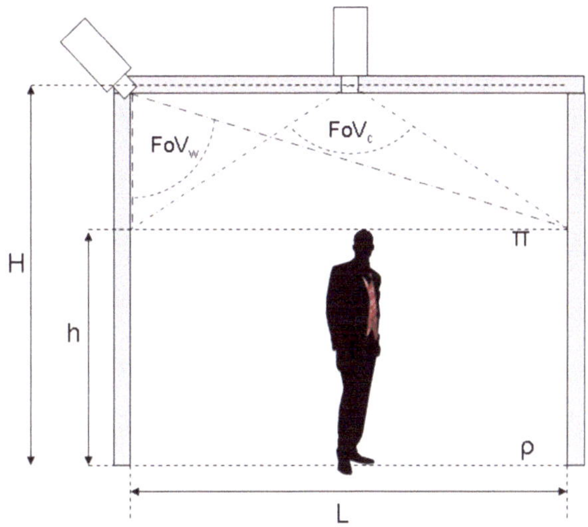

Fig. 2 Two possible camera mounting setups were considered: ceiling mounting and wall mounting

In Fig. 3 FoVc and FoVw are plotted by using Eq. 1 for three typical room having L dimensions of 3, 5 and 7 m. The distance H–h in indoor environments ranges typically from 1 to 2 m, hence a wall mounted camera requires a narrower FoV than a ceiling mounted one. The ceiling mounting configuration is less sensitive to occlusion issues, multiple reflections and flickering effects due to high reflectivity surfaces (windows, mirrors, etc.). On the other hand, in the wall mounting configuration the maximum achievable distance from the camera is greater than that achievable in the ceiling mounting configuration. However, the wall mounting configuration is more sensitive to occlusion problems and spikes may appear in the depth map due to high-reflectivity surfaces. Although ceiling mounting configuration offers many advantages, it does not allow to monitor a wide area, especially when the active sensor is positioned at a limited height from the floor plane. The previous considerations and the narrow FoV typical of TOF cameras motivate the preference of wall mounting setups in AAL contexts.

3.3 Self-Calibration of Extrinsic Parameters

Despite TOF cameras are normally intrinsically calibrated by manufacturers, the external calibration parameters must be estimated. In this section a camera self-calibration algorithm is presented, allowing to achieve a very simple installation process, in agreement with the easy-to-use feature typically required in AAL contexts. The external calibration refers to the estimation of the camera position and orientation (i.e. the camera pose) with respect to a world reference frame fixed at floor level. Both world reference frame (Ow, X, Y, Z) and camera reference frame (Oc, x, y, z) are represented in Fig. 4a in which the camera is accommodated in a wall mounting static configuration at height H from the floor plane.

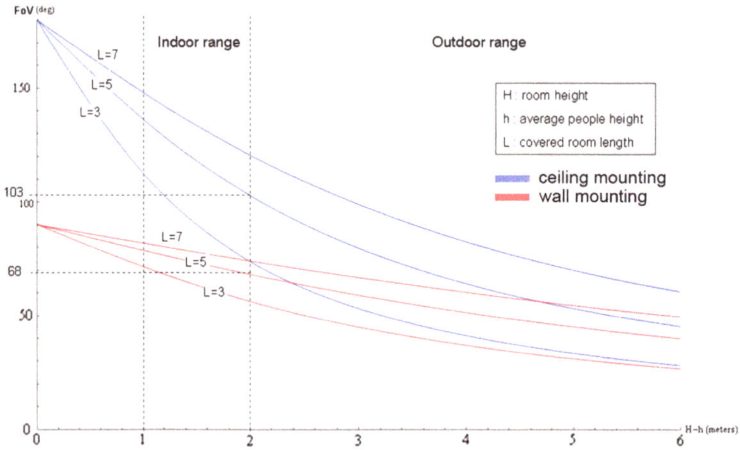

Fig. 3 The fields-of-view FoVc and FoVw are plotted by using Eq. 1 for three typical room dimensions (L=3 meters, L=5 meters and L=7 meters), in function of the distance H-h between person's head and ceiling. In indoor applications the wall mounting setup requires a narrower FoV than the ceiling mounting one

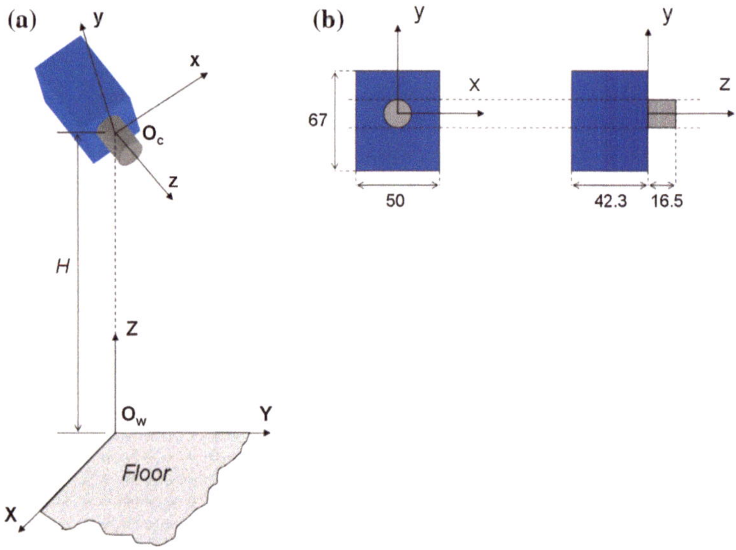

Fig. 4 (**a**) (Ow,X,Y,Z) and (Oc,x,y,z) are world and camera reference frames respectively. The camera is accommodated in a wall-mounting static setup. (**b**) Also camera dimensions are provided

In order to define the camera calibration algorithm, the following assumptions seem to be reasonable for indoor environments:

(A.1) the camera is oriented to capture a relatively large floor plane surface;

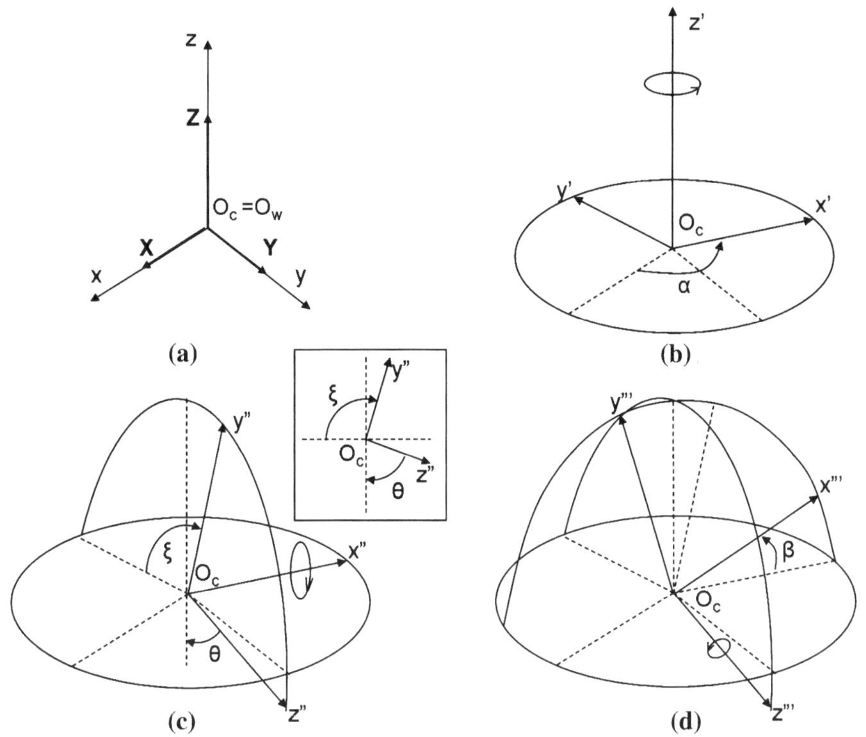

Fig. 5 The camera orientation is defined in terms of Pan (α), Tilt (θ) and Roll (β) angles by using the well known z-x-z convention. Starting from the camera reference frame aligned with the world reference one (**a**), the first rotation is performed around the z-axis of an angle α (**b**); the second rotation around the x-axis of on angle $\xi = \pi\text{-}\theta$ (**c**); and finally the third rotation around the z-axis of a β angle

(A.2) the floor plane could be covered by carpet-like surfaces;
(A.3) the presence of little objects like poufs, boxes, etc., is very limited: the floor is not entirely covered by little objects;
(A.4) the camera could capture other planar surfaces (tables, walls, etc.).

Given the previous A.1, A.2, A.3 and A.4 assumptions, a camera calibration procedure based on floor plane detection is defined. The camera orientation can be defined in terms of pan (α), tilt (θ) and roll (β) angles with respect to a world reference frame as represented in Fig. 5. Following the well known z–x–z convention the camera orientation can be represented as a composition of three rotations, starting from the world coordinated axes (Fig. 5a) and performing: (1) a rotation around the z-axis of α (Fig. 5b), (2) a rotation around the x-axis of $\xi = \pi - \theta$ (Fig. 5c), and finally (3) a rotation around the z-axis of β (Fig. 5d). In homogeneous coordinates the transformation matrix from the camera reference frame into the world reference frame can be written as follows:

$$\mathbf{M} = \begin{pmatrix} \mathbf{R} & \bar{T} \\ \bar{0}^T & 1 \end{pmatrix}, \text{ where } \bar{T} = \begin{pmatrix} 0 \\ 0 \\ H \end{pmatrix} \bar{0} = \begin{pmatrix} 0 \\ 0 \\ 0 \end{pmatrix}, \text{ and} \tag{2}$$

$$\mathbf{R} = \begin{pmatrix} \cos\alpha & -\sin\alpha & 0 \\ \sin\alpha & \cos\alpha & 0 \\ 0 & 0 & 1 \end{pmatrix} \cdot \begin{pmatrix} 1 & 0 & 0 \\ 0 & \cos\xi & -\sin\xi \\ 0 & \sin\xi & \cos\xi \end{pmatrix} \cdot \begin{pmatrix} \cos\beta & -\sin\beta & 0 \\ \sin\beta & \cos\beta & 0 \\ 0 & 0 & 1 \end{pmatrix}. \tag{3}$$

Defining the camera orientation with respect the world reference frame (that is fixed at the floor level) is the same as defining the floor plane orientation in camera coordinates. Hence, the floor plane can be written in camera coordinates by means of the Eq. 2 transforming its normal vector (0, 0, 1, 0) from world homogeneous coordinates into camera ones as follows:

$$\hat{n} = \mathbf{M}^{-1} \begin{pmatrix} 0 \\ 0 \\ 1 \\ 0 \end{pmatrix} = \begin{pmatrix} \sin\beta\sin\theta \\ \cos\beta\sin\theta \\ -\cos\theta \\ 0 \end{pmatrix}. \tag{4}$$

It is clear from the Eq. 4 that the pan angle α is irrelevant in order to define the floor plane orientation in camera reference frame. At the same conclusion one can reach from Fig. 5b since the first rotation around the z-axis of α does not change the floor plane orientation with respect the camera reference frame. The previous consideration guarantees that the only useful camera calibration parameters are (H, θ, β). In order to detect the floor plane correctly (as it will be detailed below) it is useful to express the camera parameters (H, θ, β) in terms of floor plane coefficients. Given the estimated floor plane π_F in camera coordinates:

$$\pi_F\colon a_F x + b_F y + c_F z + d_F = 0 \tag{5}$$

and its normal vector (a_F, b_F, c_F) (it is assumed that $a_F^2 + b_F^2 + c_F^2 = 1$, otherwise it can be normalized), from Eq. 4 the following relations arise:

$$\begin{cases} \sin\beta\sin\theta = a_F \\ \cos\beta\sin\theta = b_F \\ -\cos\theta = c_F \end{cases} \tag{6}$$

By using simple trigonometric considerations the camera parameters can be derived from Eqs. 5 and 6 as follows:

$$\begin{cases} \theta = \arccos(-c_F) \\ \beta = \arcsin\left(\dfrac{a_F}{\sqrt{1-c_F^2}}\right) \\ H = d_F \end{cases} \tag{7}$$

when: $0<\theta<\pi$, $-\frac{\pi}{2}\leq\beta\leq\frac{\pi}{2}$, $a_F^2+b_F^2+c_F^2=1$. Given the Eq. 5 of the estimated floor plane π_F and the person's centroid $\vec{C}=\left(c_x, c_y, c_z\right)$ in camera coordinates, the distance of $\vec{C}$ from the floor plane can be evaluated as follows:

$$h\left(\vec{C}\right) = \left|a_F c_x + b_F c_y + c_F c_z + d_F\right|. \tag{8}$$

The estimation of the external calibration parameters (θ, β, H) is accomplished during the installation of the device. Assuming that the camera is adjusted in order to look toward the floor (see A.1), the calibration plane is detected by a three-steps strategy: (1) detection; (2) filtering; (3) selection. The first step deals with the detection of enough large planes in the 3D point cloud, whereas in the second step detected planes are filtered out on the basis of some assumptions on camera orientation (defined by the given below Eq. 11). Finally, the third step selects the floor plane among all filtered planes. The planes detection algorithm searches iteratively the largest plane in the 3D point cloud removing points belonging to the detected plane at each iteration, as explained by the pseudo-code in Algorithm 1.

Algorithm 1 Self-calibration **Algorithm**: Planes detection

```
S = {(x_k,y_k,z_k):k=1,…,25344}    # 3D point cloud
N_0 = |S|                          # cardinality of S
L_S = {}
L_Π = {}
i=1
repeat
  S_i is the largest subset in S fitting Π_i with Ransac
  Π_i=(a_i,b_i,c_i,d_i)  # parameters of the Ransac fitted plane
  S ← S\S_i
  L_S ← L_S U {S_i}
  L_Π ← L_Π U { Π_i}
  i ← i+1
until (|S|/N_0 < p)
```

Hence at a given iteration, the algorithm works with a subset of the 3D points used in the previous iteration. The detection procedure finishes when the size of the subset is lower than a prefixed percentage p (greater than 30) of the starting points. Since measured distances are normally affected by noise, planes are detected by using a RANSAC-based approach [32] which is robust to outliers. Let the i-th iteration of the algorithm, the RANSAC plane detector provides four parameters (a_i, b_i, c_i, d_i) describing the implicit model of the i-th fitted plane π_i in camera coordinates:

$$a_i p_x + b_i p_y + c_i p_z + d_i = 0, \tag{9}$$

with $a_i^2 + b_i^2 + c_i^2 = 1$, for each point $P = (p_x, p_y, p_z)$ belonging to the detected plane π_i. For each detected plane π_i the camera tilt and roll angles (θ_i, β_i) are evaluated by using Eqs. 7 and 9:

$$\begin{cases} \theta_i = \arccos(-c_i) \\ \beta_i = \arcsin\left(\dfrac{a_i}{\sqrt{1 - c_i^2}}\right) \end{cases} \tag{10}$$

The (θ_i, β_i) angles are used in the second step to filter out planes not satisfying the following constraints:

$$\begin{cases} -20^\circ \le \beta_i \le 20^\circ \\ 23.75^\circ \le \theta_i \le 66.25^\circ \end{cases} \tag{11}$$

Since not only the floor plane satisfies Eq. 11 but even all coplanar planes, the floor plane is selected as the farthest plane from the camera. Therefore, in the third algorithmic step the floor plane π_F is selected such that the subscript index F is:

$$F = \underset{1 \le i \le m}{\arg\max} \left\{ |d_i| : a_i^2 + b_i^2 + c_i^2 = 1 \right\}. \tag{12}$$

The self-calibration procedure was validated by using a MEMS-based Inertial Measurement Unit (IMU) [33] and a Laser Measurement System (LMS) both attached to the 3D range camera in order to derive ground truth data. The IMU sensor provided drift-free 3D orientation with a static accuracy better than 0.5°, whereas the LMS measures distances with accuracy of ± 1.0 mm. The calibration procedure was evaluated in several typical household environments such as living room, kitchen, bedroom, corridor and bathroom, and varying the following parameters:

(P.1) the percentage of floor occupancy by using three groups of objects: (a) carpet-like surfaces with thickness no greater than 5 cm, (b) furniture with height greater than 50 cm (like chairs, beds, nightstands, etc.), and (c) little objects (like poufs) having height ranging from 10 to 30 cm;
(P.2) the camera height from the floor plane, ranging from 2.00 to 2.70 m;
(P.3) the camera orientation β and θ angles, with $-20^\circ \le \beta \le 20^\circ$ and $23.75^\circ \le \theta \le 66.25^\circ$.

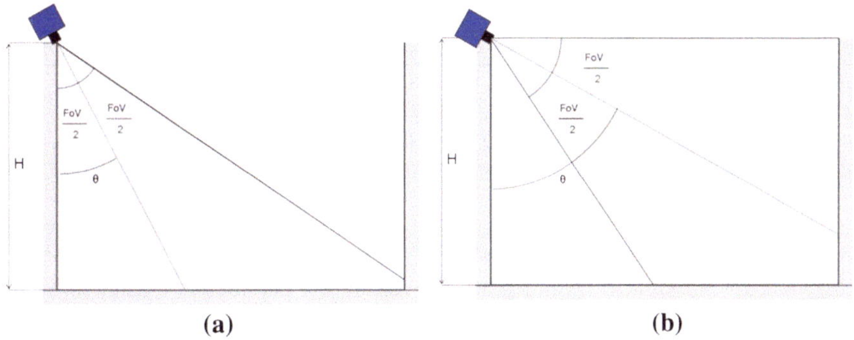

Fig. 6 The geometries involved into the definition of θ lower bound and upper bound are reported in **a**) and **b**) respectively. In order to avoid that camera captures same portion of wall the θ must be greater than FoV/2 that is 23.75°. On the other hand, given that in surveillance applications it is not useful camera captures more than 2.00 m on the opposite wall, the maximum useful value for θ is 90°-FoV/2 that is 66.25°

The camera height range was defined considering that normally the ceiling height is lower than 2.70 m and it is not recommended to install camera at a height lower than 2.00 m (to prevent both safety problem and saturation effects). The range for β seems to be reasonable, since in surveillance application usually one tries to accommodate camera without strong roll rotations. The maximum camera FoV is 47.5°, hence it was not useful to wall mount the camera with tilt angle lower than FoV/2 = 23.75° in order to prevent wall being captured by camera. The geometry involved into definition of θ lower bound is shown in Fig. 6a. Moreover, the maximum value of θ angle was defined considering that normally in surveillance applications it is not useful camera captures more than 2.00 m on the opposite wall and for this reason the maximum useful value for θ is 90°−FoV/2 = 66.25° as shown in Fig. 6b. Camera orientation and position values indicated by the previously mentioned P.2 and P.3 parameters allow to monitor virtually any floor portion inside a typical sized household room of about 4 × 4 m. The values of (H, θ, β) taken into account during the validation of the self-calibration procedure are summarized in Table 4 for a total amount of 324 camera configurations.

Since the measurement of people movements could be demoted by errors in camera calibration, the performance of self-calibration algorithm was evaluated. The calibration procedure was validated considering the relative errors E_H, E_θ and E_β defined as follows:

$$E_H = \frac{|H - \hat{H}|}{H}, \quad E_\theta = \frac{|\theta - \hat{\theta}|}{\theta}, \quad E_\beta = \frac{|\beta - \hat{\beta}|}{\beta}, \tag{13}$$

Table 4 Camera calibration parameters. Values used during tests

Parameter	Tested values
Height, H (m)	2.00, 2.14, 2.28, 2.42, 2.56, 2.70
Tilt angle, θ (deg)	23.75, 32.25, 40.75, 49.25, 57.75, 66.25
Roll angle, β (deg)	−20.00, −15.00, −10.00, −5.00, 0, 5.00, 10.00, 15.00, 20.00

where H, θ and β are the calibration parameters measured with camera attached IMU and LMS, while $\hat{E}$, $\hat{\theta}$ and $\hat{\beta}$ are the calibration parameters estimated by the self-calibration algorithm. Furthermore, the precision of the calibration procedure was evaluated for different percentage of available floor surface. The environments were arranged in order to obtain several percentage of floor occupancy considering both uncovered and covered floor with carpet surfaces in percentages ranging from 20 to 80 %.

3.4 Background Modeling, People Segmentation and Tracking

All kind of vision-based recognition applications require that some pre-processing tasks are performed before that features are extracted. At this purpose a well-established framework is adopted including early vision algorithms for background modeling, foreground segmentation and people tracking. Since the adopted pre-processing framework is based on well-known concepts, only practical implementation details are given in this subsection omitting further theoretical details. Interested readers in pre-processing aspects can refer to [34] for further details. A Bayesian segmentation is used to detect the 3D elderly silhouette in depth images. In order to perform an automatic foreground extraction, an improved version [35] of the method proposed by Stauffer and Grimson [36] (Mixture of Gaussians—MoGs method) has been enhanced by considering depth information. In the traditional formulation of MoGs method, the probability that a pixel belongs to the foreground is modeled as a sum of K normal functions. For each pixel some of these Gaussians model the background and the others the foreground. In the suggested formulation, at each frame the model parameters are updated by using the Expectation Maximization (EM) algorithm and a pixel is considered to belong to the foreground if its depth does not belong to any of the Gaussians. The EM algorithm allows to update the Gaussian parameters according to a fixed learning rate that controls the adaptation speed. The well-known problem of this approach is the balancing between model convergence speed and stability. The MoGs scheme improves the convergence rate without compromising model stability: the background model is updated online and the global static retention factor of the traditional formulation is replaced with an adaptive learning rate calculated for each Gaussian at every frame. The segmentation involves a binary classification problem based on P(B|z), where z is the depth value at time t and B the background class. According to the background model process, let $g(z; \mu_k; \sigma_k)$ the probability that a particular depth belongs to the k-th Gaussian function G_k having weight ω_k ($P(G_k) = \omega_k$). With an explicit representation of the distribution P(z) as a mixture:

$$P(z) = \sum_{k=1}^{K} P(G_k)P(z|G_k) = \sum_{k=1}^{K} \omega_k \cdot g(z; \mu_k, \sigma_k) \tag{14}$$

the posterior probability can be expressed in terms of the mixture components $P(G_k)$ and $P(z|G_k)$. Therefore, by using the Bayes rule the density $P(B|G_k)$ can be expressed as:

$$P(B|z) = \sum_{k=1}^{K} P(B|G_k)P(G_k|z) = \frac{\sum_{k=1}^{K} P(z|G_k)P(G_k)P(B|G_k)}{\sum_{k=1}^{K} P(z|G_k)P(G_k)} \tag{15}$$

To estimate $P(B|G_k)$ a sigmoid function on ω/σ is trained using the logistic regression:

$$\hat{P}(B|G_k) = f\left(\frac{\omega_k}{\sigma_k}; a, b\right) = \frac{1}{1 + e^{-a\frac{\omega_k}{\sigma_k}+b}} \tag{16}$$

with a = 96 and b = 3, evaluated by training. Once P(z) and $P(B|G_k)$ are estimated, foreground regions are those for which the relation $P(B|z) < 0.5$ is satisfied. The default threshold 0.5 worked quite well with a fitted sigmoid trained on representative data.

The whole segmentation process was implemented in C++ with the support of OpenCv library [37], guarantying real-time functioning. Once person's silhouette has been detected and its centroid (i.e., approximately near the center-of-mass) has been estimated, a tracking strategy allows to link people silhouettes in different time instants. A widely used approach for tracking is the Kalman filter [38] applied to each segmented object. This approach requires a high complexity management system to deal with the multiple hypotheses necessary to track objects. Due to the non-linear nature of human motion, a stochastic approach is used based on the ConDensation scheme (Conditional Density Propagation over time [39]) that is able to perform tracking with multiple hypotheses directly in range images (500 samples are used for people tracking). The 3D centroid is predicted frame-by-frame in range data, according to a state vector defined by merging position and velocity vectors of the centroid. The tracker is realized by thresholding the Euclidean distance between the predicted centroid position and its measured version in the adjacent time step. As discussed for the segmentation step, the ConDensation algorithm implementation in OpenCv library was used allowing to exploit the advantages of a low-level implementation.

3.5 The Fall Detection Strategy

A fall event is detected when the following events happen:

(1) the distance of the person's centre-of-mass (approximated with the silhouette's centroid) with respect the floor plane decreases below to 0.40 m within a time window of about 900 ms;
(2) the people silhouette movements remain negligible within a time window of about 4 s.

The centroid position vector over time $\vec{C}(t) = \left(c_x(t), c_y(t), c_z(t)\right)$ is estimated from range images by using the previously described algorithms for segmentation and tracking. The distance $h(t)$ of the centroid from the floor plane is equal to the z coordinate of the centroid in the world reference frame and hence it can be calculated by using the Eq. 8 as follows:

$$h(t) = \sin\beta \sin\theta \cdot c_x(t) + \cos\beta \sin\theta \cdot c_y(t) - \cos\theta \cdot c_z(t) + H \quad (17)$$

where (θ, β, H) are the previously described calibration. The proposed scheme for fall detection works when a whole human silhouette is detected and also when a partial occlusion occurs. The centroid height estimation was validated by using a test object with known height of 0.40 m and accommodated in nine different positions within a surface of 4 × 4 m as shown in Fig. 7. The height measurements were repeated for each camera position and orientation value reported in the previously discussed Table 4. For each measurement the following quantity was evaluated:

$$\Delta h = \left|\hat{h} - 0.4\right| \quad (18)$$

where $\hat{h}$ is the estimated height. The implemented fall detector is able to process range data real-time (up to 25 fps). Fall-detection performance was evaluated by

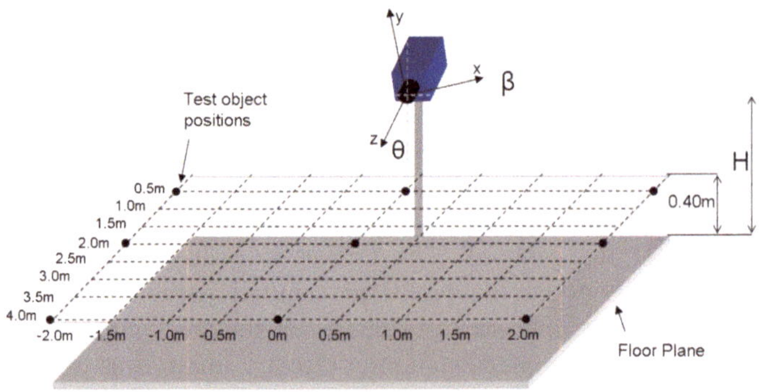

Fig. 7 Accuracy and precision of centroid height estimation by 3D camera was validated by using a test object with known height of 0.40 m and accommodated in 9 different positions within a surface of 4 × 4 m

using data collected during the simulation of falls in real-home scenarios such as living room, kitchen, bed room and bathroom. The performance of the overall system is quantified as suggested by Noury et al. [22] by using sensitivity and specificity measures, defined as follows:

$$\text{Sensitivity} = \text{True Positives} \,/\, (\text{True Positives} + \text{False Negatives}), \quad (19)$$

$$\text{Specificity} = \text{True Negatives} \,/\, (\text{True Negatives} + \text{False Positives}). \quad (20)$$

3.6 The Simulation Setup

The simulation of realistic fall events was performed with the involvement of 13 stuntmen. All participants were healthy male, from 30 to 40 years old and height between 1.55 and 1.95 m. A total amount of 460 actions were simulated of which 260 were falls in all directions (backward, forward and lateral) and with/without recovery post fall. The simulated falls were compliant with those categorized by Noury et al. [40] and they can be grouped into the following seven categories:

(F1) backward fall ending lying (FBRS),
(F2) backward fall ending lying in lateral position (FBRL),
(F3) backward fall with recovery (FBWR),
(F4) forward fall with forward arm protection (FFRA),
(F5) forward fall ending lying flat (FFRS),
(F6) forward fall with recovery (FFWR),
(F7) lateral fall (FL).

Each participant was involved in two sequences simulating ten falls (one for each type from F1 to F6 and four times the type F7) for each sequence. Since the falls in the lateral direction are associated with a high risk of hip fractures in elderly people [41], the simulation of this type of fall (F7) was mainly stressed. Moreover, in order to stress the reliability of the framework, the fall detector was validated in presence of occluding objects; each participant performed at least one half of falls (i.e. five falls in each sequence) occluded by a table, a chair or a sofa.

3.7 Experimental Results

3.7.1 Floor Detection

The precision of the self-calibration procedure was evaluated by repeating the procedure at increasing percentage of available floor both with and without carpet covering. Results without carpet are reported in Table 5.

Table 5 Mean and standard deviation of E_H, E_θ and E_β for percentage of available floor

Floor occupancy	E_H		E_θ		E_β		Δh (m)	
%	Mean	Std.dev.	Mean	Std.dev.	Mean	Std.dev.	Mean	Std.dev.
20.00	0.4070	0.0302	0.1324	0.0311	0.0232	0.0072	0.3196	0.0456
30.00	0.0209	0.0027	0.0181	0.0035	0.0180	0.0029	0.0502	0.0034
40.00	0.0189	0.0028	0.0167	0.0038	0.0151	0.0030	0.0420	0.0030
50.00	0.0150	0.0028	0.0154	0.0039	0.0128	0.0028	0.0414	0.0028
60.00	0.0142	0.0033	0.0131	0.0041	0.0130	0.0029	0.0228	0.0024

Table 6 Mean and standard deviation of E_H, E_θ and E_β at different percentages of available floor and different percentages of carpet covering

Floor Occ (%)	Carpet Occ (%)	E_H		E_θ		E_β		Δh (m)	
		Mean	Std dev	Mean	Std dev	Mean	Std dev	Mean	Std dev
30.00	20.00	0.3484	0.0298	0.1159	0.0223	0.0219	0.0050	0.1256	0.0085
30.00	40.00	0.3477	0.0308	0.1167	0.0222	0.0232	0.0055	0.1244	0.0068
30.00	60.00	0.3490	0.0305	0.1155	0.0216	0.0213	0.0049	0.1080	0.0056
30.00	80.00	0.3480	0.0300	0.1172	0.0221	0.0227	0.0055	0.0858	0.0045
40.00	20.00	0.0208	0.0028	0.0187	0.0033	0.0176	0.0027	0.0570	0.0033
40.00	40.00	0.0216	0.0030	0.0176	0.0037	0.0175	0.0037	0.0534	0.0049
40.00	60.00	0.0199	0.0018	0.0185	0.0032	0.0182	0.0026	0.0596	0.0038
40.00	80.00	0.0209	0.0016	0.0177	0.0035	0.0172	0.0028	0.0542	0.0028
50.00	20.00	0.0187	0.0025	0.0167	0.0038	0.0147	0.0030	0.0522	0.0026
50.00	40.00	0.0194	0.0033	0.0161	0.0045	0.0154	0.0039	0.0560	0.0021
50.00	60.00	0.0187	0.0031	0.0178	0.0037	0.0153	0.0038	0.0380	0.0007
50.00	80.00	0.0193	0.0029	0.0177	0.0041	0.0140	0.0029	0.0218	0.0009

When the percentage of available floor was greater than 30 % the relative errors E_H, E_θ and E_β were less than 2 % with uncertainty less than 1.23 % (according to the 3σ rule in the theory of errors), whereas the measure inaccuracy of the centroid height was less than 5.0 cm and its uncertainty was less than 10.2 mm. The calibration procedure was evaluated also with carpet-like surfaces covering partially the floor plane. Mean and standard deviation of relative errors reported in Table 6 were estimated in correspondence of variable percentages of available floor surface (not occupied by furniture) and carpet-like surfaces. In the worst case in which many carpets were arranged in various positions and with long pile thickness (near to 5 cm) it was needed a percentage of available planar surface at floor level greater than 40 % in order to relative errors E_H, E_θ and E_β went below the 2 % with an uncertainty less than 1.35 %. In the same situation the measure inaccuracy of the centroid height was less than 6.0 cm with uncertainty less than 14.7 mm.

Some range images used by self-calibration algorithm during data collection in typical dwelling rooms are shown in Fig. 8. Column (a) reports intensity images

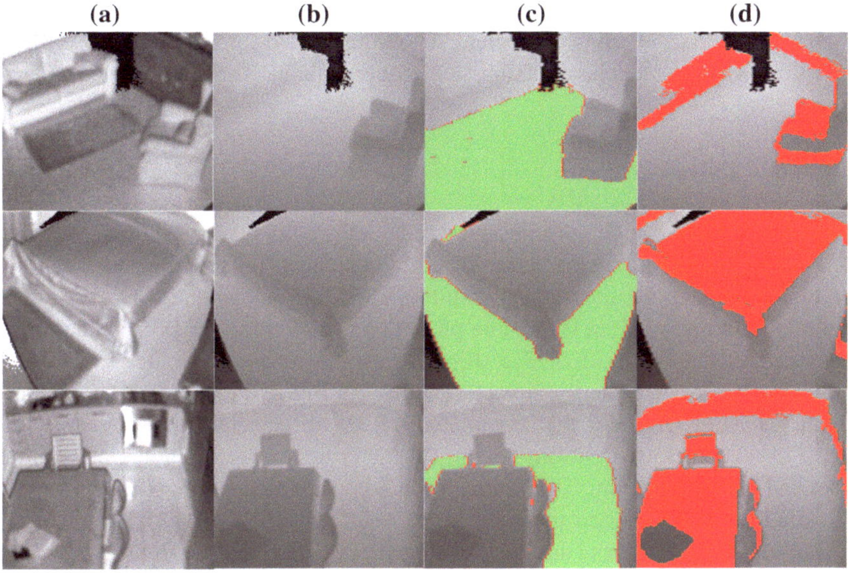

Fig. 8 Some range images used by self-calibration algorithm during data collection in typical dwelling rooms. Figure shows intensity images (column **a**), range images (column **b**), floor planes correctly detected in range images (column **c**) and rejected planes in range images (column **d**)

whereas column (b) reports corresponding range images. Floor planes correctly identified by self-calibration algorithm are shown in column (c), whereas rejected planes are shown in column (d). The first two rows in the figure are referred to rooms with a carpet as it is visible by corresponding intensity images. However, the carpet was not detected by 3D camera since its thickness of about 0.5 cm was lower than the maximum accuracy achievable at the distance of 4 m.

3.7.2 People Detection and Tracking

The position of the people centroid is estimated from the segmented silhouette in the range image and its evolution over time is pursued by using the tracking algorithm detailed in the previous section. Segmentation results are illustrated, as anticipated in the previous section, in Fig. 1 by reporting a critical situation in which passive vision is effected by camouflage effects. Since range camera is not sensitive to illumination or shadows, the elderly silhouette has been accurately segmented from range images (Fig. 1b). In order to emphasize the goodness of range images for segmentation, the same segmentation scheme has been applied to intensity images and the corresponding result is shown (Fig. 1d): the poor quality of the segmentation is due to both the inability of the system to model the background and the presence of camouflage effects (the garment presents chromatic information similar to the wall at the back). The presented results are obtained by using K = 3 Gaussian functions in the background modeling process with $\alpha = 0.005$ as learning coefficient. The time needed for segmentation is about of 15 ms whereas the classification step requires about 10 ms.

3.7.3 Fall Detection

The 3D centroid height profile over time allows to distinct falls from other activities by thresholding the centroid height and unmoving time interval as shown in Fig. 9. In Fig. 10 the typical trend of the centroid height during a fall is reported. During a fall, at least three phases can be distinguished [22]: the pre fall phase (indicated with I0 in Fig. 10), the critical phase (indicated with I1 in Fig. 10), the post fall phase (indicated with I2 in Fig. 10) and the recovery phase (indicated with I3 in Fig. 10). Simulated falls were detected by using three features: (1) the person's centroid height, (2) the critical phase duration, and (3) the post fall phase duration. The three thresholds identified during the analysis of recorded falls are reported in Table 7.

Fall-detection performance was evaluated by using the previously described dataset of simulated falls, with and without the presence of occluding objects such as tables, chairs, sofas, etc. Firstly, results without occlusions will be presented in the following. The first threshold TH1 alone was able to detect correctly all simulated falls achieving a sensitivity of 100 %, although it was not able to distinguish between a fall and a "fall with recovery" or between a fall and a "voluntary lying down on floor". A statistical visualization of results related to the threshold TH1 is shown in Fig. 11. The threshold TH1 alone correctly identified

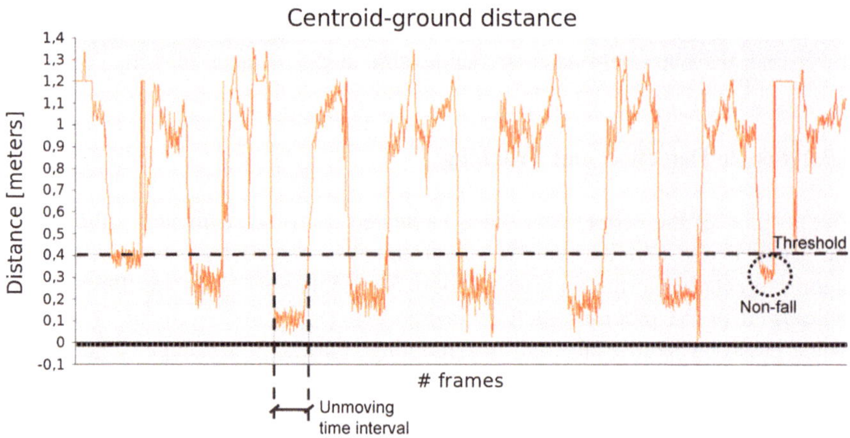

Fig. 9 Centroid height trend analyzed in order to detect falls. The first seven events are correctly classified as fall, whereas the last one is correctly classified as a non-fall since the unmoving duration is too short

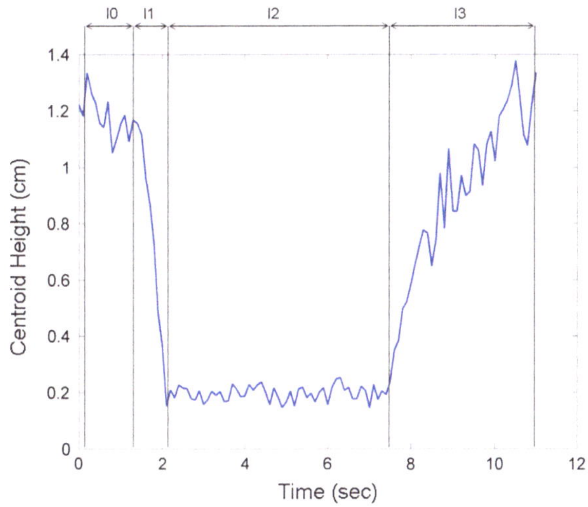

Fig. 10 The typical trend of the centroid height during a fall is shown. During a fall can be distinguished the following phases: the pre fall phase (I0), the critical phase (I1), the post fall phase (I2) and the recovery phase (I3)

63.5 % of ADLs as non-falls achieving a specificity of 63.5 %. By adding the second threshold TH2 a specificity of 79.4 % was obtained, since the threshold TH2 allowed to discriminate correctly a "voluntary lying down on floor" from an involuntary fall characterized by a shorter duration of the critical phase. The statistical visualization of TH2 discrimination capability is shown in Fig. 12. By using the TH1, TH2 and TH3 thresholds simultaneously a specificity of 100 % was achieved, since the threshold TH3 allowed to detect correctly falls with recovery as non-falls by considering the duration of the post fall phase shorter than 4 s in case of recovery. Conversely, in presence of occluding objects it was not possible to detect correctly all simulated falls. Although partial occluded falls happened behind a small object such as a chair were correctly handled, others seriously occluded falls such as those occurred behind a large table were prone to generate false negatives. Similarly, simulated falls with recovery gave rise to false positives due to the impossibility to detect occluded post fall movements. By using the three thresholds TH1, TH2 and TH3 defined in Table 7 a specificity of 97.3 % and a sensitivity of 80.0 % where obtained when falls were occluded by furniture. The previously discussed fall-detection performance is summarized in the following Table 8.

Table 7 Fall-detection threshold values

Threshold	TH1	TH2	TH3
Measure	Centroid height	Critical phase duration	Post fall phase duration
Unit	Meters	Milliseconds	Seconds
Value	0.40	900	4

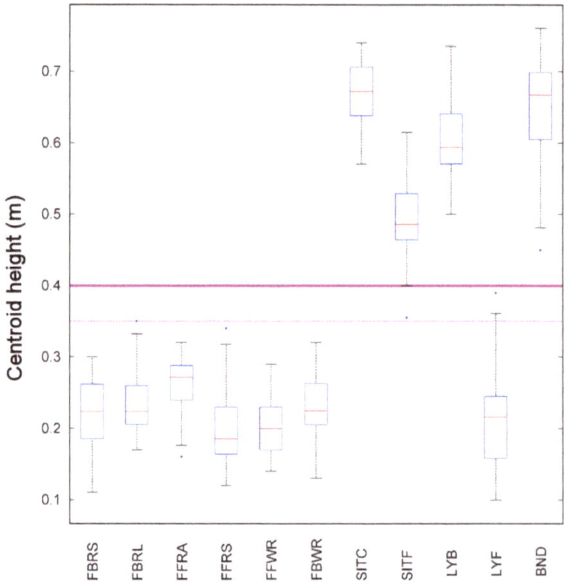

Fig. 11 Statistical visualization with boxplot of minimum centroid height value during falls and ADLs. The threshold TH1 alone correctly identified SITC, SITF, LYB and BND as non-falls, but it was unable to distinguish falls, falls with recovery (FFWR, FBWR) and voluntary "lying down on floor" (LYF)

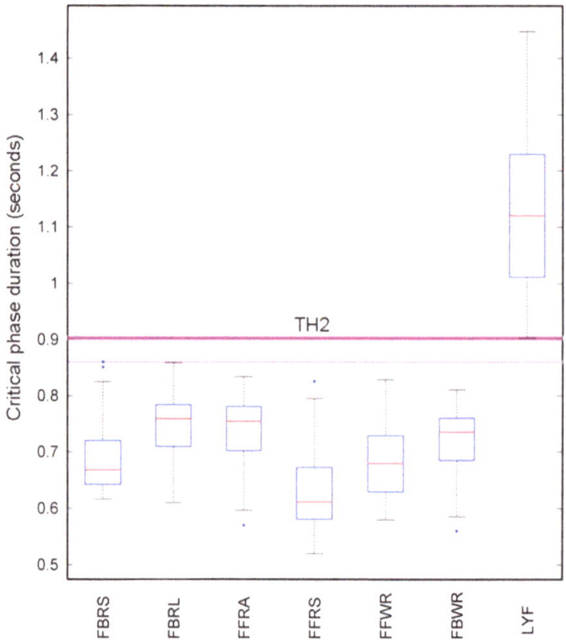

Fig. 12 Statistical visualization with boxplot of critical phase duration during falls. The threshold TH2 allowed to discriminate correctly a voluntary "lying down on floor" (LYF) from an involuntary fall characterized by a shorter duration of the critical phase

Table 8 Fall-detection performance

Thresholds	Sensitivity		Specificity	
	Without occlusions (%)	With partial occlusions	Without occlusions (%)	With partial occlusions
TH1	100	–	63.5	–
TH1, TH2	100	–	79.4	–
TH1, TH2, TH3	100	80.0 %	100	97.3 %

4 A TOF Camera-Based Framework for Posture Recognition

The human posture analysis is a highly active research area in computer vision, dealing with the ill-posed problem of inferring the pose and motion of a highly articulated and self-occluding non-rigid 3D object (a human body) from images. Traditionally, posture analysis algorithms are categorized on whether a body model is employed (either directly or indirectly) or not [42]. Model-based techniques use a priori information about human body shape in order to reconstruct the entire posture kinematics. Within this approach the body is usually represented with a stochastic region model or a stick figure model [4, 43]. Model-based approaches are quite expensive in term of computational resources and they are generally well suited for human motion capture in which the motion of significant segments of the human body must be tracked (i.e. head, arms and legs). On the other hand, model-free techniques estimate body posture directly on individual images without any preliminary information about the body shape and so allowing to overcome limitations of tracking features over long sequences [44, 45]. Within the model-free category two different approaches are investigated in literature: the probabilistic assemblies of parts and the example-based methods. In the first approach individual body parts are first detected and then assembled to infer the body posture [46]; whereas the second approach directly learns the mapping from image space to 3D body space [47]. Concerning the vision system, monocular and stereo/multiple view systems are the most investigated in literature of human posture recognition. Here similar considerations previously made for fall detection apply. Posture estimation from a monocular view is considerably more difficult than estimation from stereo/multiple cameras since a single view image can be strongly affected by perspective ambiguity making troublesome to correctly discriminate posture from unfavorable point-of-view [4, 5]. Furthermore, about one third of all DOFs are almost unobservable by using a monocular camera system [47]. Stereo and multiple view vision systems overcome perspective problems exploiting 3D geometric representation of human shape [48, 49]. However, stereo and multiple view vision are affected by many drawbacks discussed in the previous section concerning vision systems used in fall detection literature. Hence, the adoption of TOF vision is increasingly investigated also for human posture recognition [8–11, 15]. TOF cameras provide dense depth measurements at every point in the scene at high frame rates, allowing to disambiguate poses with similar appearance that can confuse monocular systems or overload stereo/multiple view systems due to the correspondence search between two or more views.

In this section, two different feature extraction approaches are presented, satisfying different requirements exhibited by AAL applications. In fact, gathered posture details and operational distance from the camera are usually inversely proportional: rehabilitation exercises can be performed at few meters from the camera (e.g., less than 3 m when the camera is upper pose on a television screen) and many postural details are required in order to check the correctness of exercise execution; while, conversely, critical events can occur at a greater distance from the camera but few postural detail are sufficient for critical event detection. This motivates the investigation of two feature extraction approaches having different discrimination capabilities in terms of gathered human postural information. The first is a topological approach in which the Morse theory is exploited in order to extract a Reeb graph-based skeleton representation of the human body [15]. A high level of detail can be achieved within a distance from camera up to 4 m. On the other hand, the second feature extraction approach advantages execution speed against details pursing a volumetric strategy based on the analysis of the 3D spatial distribution of the human body [50]. The discrimination capabilities of the two feature extraction approaches are evaluated by using a statistical learning methodology and compared on the basis of a common dataset of four basic human postures: standing, bent, sitting and lying down.

The pre-processing framework (background modeling, foreground segmentation, people tracking, and so on) is the same of those presented in the previous section devoted to detection of falls. Moreover, the same considerations apply for camera mounting setup and the self-calibration procedure. Instead, the TOF camera is the new MESA SR4000. Several improvements put the MESA SR4000 at the state-of-the-art in the field of TOF RIM over the previous MESA SR3000; for example the new camera is full noiseless (0 db), the power consumption has been reduced of about 50 % under normal operation, two highly accurate (manufacturer recommended) non-ambiguity ranges are now available (5.0 m at 30 MHz, and 10 m at 15 MHz) instead of one (7.5 m at 20 MHz), the camera focus is now adjustable in order to obtain accurate 3D data whereas the SR3000 does not have adjustable focus, only to cite a few of them. Further details on cameras comparison can be found in the comparative sheet provided by the manufacturer [31].

4.1 The Experimental Setup

Four main postures, standing, sitting, bent and lying down were simulated during basic Activities of Daily Living (ADLs) involving the interaction with common objects (i.e., tables, sofas, chairs, room/furniture doors, kitchen units, etc.) in order to evaluate the reliability of extracted features in typical home environments. The four main postures are simulated during the following five basic ADLs:

(A1) sitting down on a chair (height, 47 cm) and then stand up (SITC),
(A2) sitting down on floor and then stand up (SITF),
(A3) lying down on a bed (height, 52 cm) and then stand up (LYB),
(A4) lying down on floor and then stand up (LYF),
(A5) bent down to catch something on the floor (BND).

Postures simulation involved only ten subjects among the 13 previously said, ranging in age from 35 to 40 years and height from 1.72 to 1.95. Each subject performed four times every action from A1 to A5 for a total amount of 200 simulated tasks chosen from those actions that mainly might be confused with falls. Other actions in addition to those listed from A1 to A5 were performed such as walking around and dropping objects on floor.

4.2 The Topological Approach

The topological features describe the human posture at a high level of detail; many body segments can be discriminated such as head, trunk, arms and legs, exploiting the full potential offered by range imaging. The intrinsic topology of a generic shape, such as a human body scan captured by a range camera, can be encoded in a graph as suggested by Werghi et al. [51]. A Reeb graph represents the hierarchical evolution of level-set curves on a manifold (that is a mathematical object more general than a classic surface) providing a powerful tool to understand intrinsic topology of any shape [52]. Moreover, defined a real-valued function on a manifold, the Reeb graph nodes represent the level-set curves of the function on the manifold. This function is called Morse function if it has no degenerated critical points on the manifold. Several Morse functions can be defined of which a few are depicted in Fig. 13. The directional height function is shown in Fig. 13a, b along horizontal and oblique directions, respectively. The radial distance function is shown in Fig. 13b and the geodesic distance function in Fig. 13d. For each function the respective level-set curves are highlighted with white colored lines or curves. In recognition applications the main aspect related to the Reeb graph concerns the invariance property under some transformations such as scale and rotation. Among all Morse functions reported in Fig. 13 only the radial distance and the geodesic distance are invariant under affine (translation, scale, rotation) transformations. In addition, the geodesic distance function is invariant under isometric transformations, i.e. those transformations that preserve the length of the path joining two generic points. The isometric invariance is very useful for posture recognition as shown in Fig. 13e in which the path joining the centroid C with the silhouette's left hand remains of the same length after a postural change. Furthermore, geodesic distance function allows to exploit the full potential offered by range imaging since it can be defined on a 3D mesh surface. Otherwise, by using

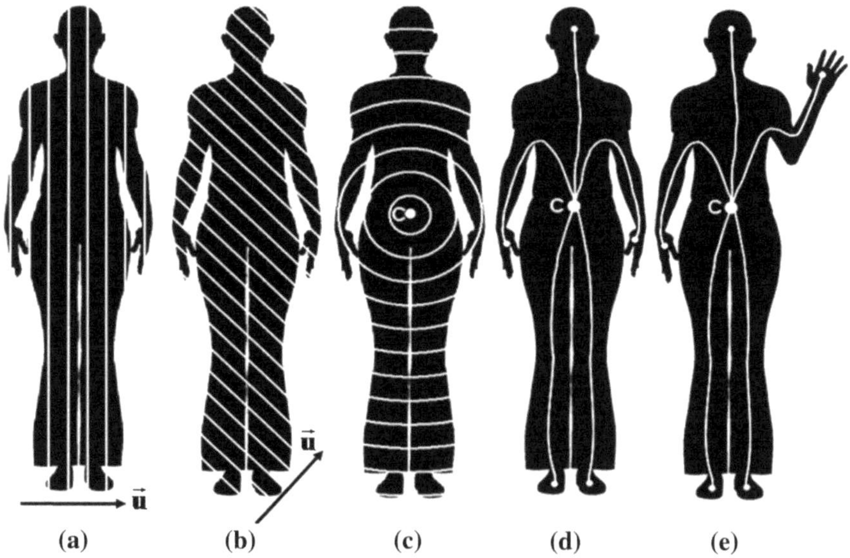

Fig. 13 Three Morse functions: Directional Height Function along the **a**) horizontal and **b**) oblique directions with level-sets depicted in white; **c**) Radial Distance Function with level-sets in white; **d**) Geodesic Distance Function with length paths in white and **e**) the same Geodesic Distance Function under a postural change

monocular passive vision the geodesic distance map is defined on a flatten foreground that can be affected by self-occlusions. This situation is depicted in Fig. 14, in which the geodesic distance map of Fig. 14c is defined on the flatten foreground of Fig. 14b obtained segmenting the original image in Fig. 14a: the mannequin's left hand results confused with the body trunk in the geodesic distance map. On the other hand, by using the range image shown in Fig. 14d a geodesic distance map can be computed without perspective ambiguity as shown in Fig. 14e.

Therefore, in this study the Reeb graph is extracted by using the geodesic distance function as described in the following. Given a range image, the Reeb function is defined as $f = G(x, y)$ with $(x, y) \in I$, where G is the geodesic distance map generated starting from the range image and I is the segmented image region. The geodesic distance map is generated in a two steps procedure: a connected mesh is computed from the range image and then geodesic distances are estimated by using the well-known Dijkstra algorithm on the connected mesh [53]. In Fig. 15c the geodesic map related to depth map of Fig. 15a is reported. Colors represent the distance of each surface point from the starting point (dark blue region): nearest points are blue, farthest ones are red. Whereas, Fig. 15b reports the connected mesh from which the geodesic map is computed.

Hence, starting from the geodesic-based Morse function (i.e., the geodesic map) the Reeb graph is extracted according to the methodology suggested by Werghi et al. [54]. Firstly, the co-domain of the real-valued Morse function f is subdivided in regular intervals as follows:

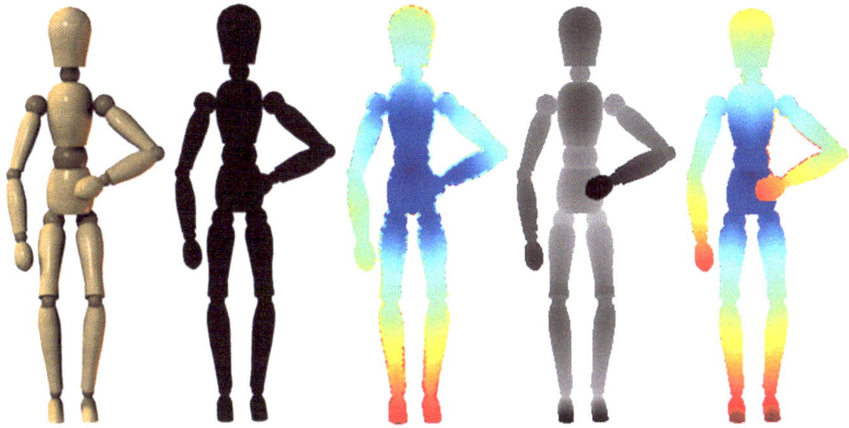

Fig. 14 Starting from the original image **a**), geodesic distance maps in **c**) and **e**) are computed from the flatten foreground **b**) and the range image **d**), respectively. Grayscale levels in the range image **d**) represent the distance of each point from camera: nearest points are dark, farthest are light. Colors in geodesic maps **c**) and **e**) represent the distance of each point from the starting point (dark blue region): nearest points are blue, farthest ones are red

$$z_0 = \min_{(x,y)\in I} f(x,y),\ z_N = \max_{(x,y)\in I} f(x,y),\ z_k = z_0 + k\frac{z_N - z_0}{N},\ \forall k \in \{0,\ldots,N\} \quad (21)$$

Then, $\Re_k$ support regions and S_k level-sets are defined, at the previously fixed intervals, as follows:

$$\Re_k = \{(x,y) \in I | z_k \leq f(x,y) < z_{k+1}\},\ S_k = f(\Re_k),\ \forall k \in \{0,\ldots,N\}. \quad (22)$$

The Reeb graph is obtained by associating each level-set S_k to a graph node and linking together two graph nodes when the corresponding support regions are connected. More precisely, two support regions $\Re_k$ and $\Re_i$ are connected if the following condition is satisfied:

$$\exists(x_k,y_k) \in \Re_k, \exists(x_i,y_i) \in \Re_i \ \ni' \ \|P_k - P_i\| \leq d, \quad (23)$$

$$P_i = \begin{bmatrix} X(x_i,y_i) \\ Y(x_i,y_i) \\ Z(x_i,y_i) \end{bmatrix},\ P_k = \begin{bmatrix} X(x_k,y_k) \\ Y(x_k,y_k) \\ Z(x_k,y_k) \end{bmatrix} \quad (24)$$

where $\|\cdot\|$ denotes the Euclidean distance between points P_i and P_k, whereas $X(\cdot,\cdot)$, $Y(\cdot,\cdot)$, $Z(\cdot,\cdot)$ are the world coordinates of each range image point indexed by $(x,y) \in I$ and d is a threshold defined according to the maximum distance between connected points and depends on the choice of N in Eq. 21. Figure 15d reports the Reeb graph related to the range image shown in Fig. 15a.

In order to define the feature vector, the Reeb graph is inspected looking for the graph nodes having the greater degree (i.e. the number of edges incident on a

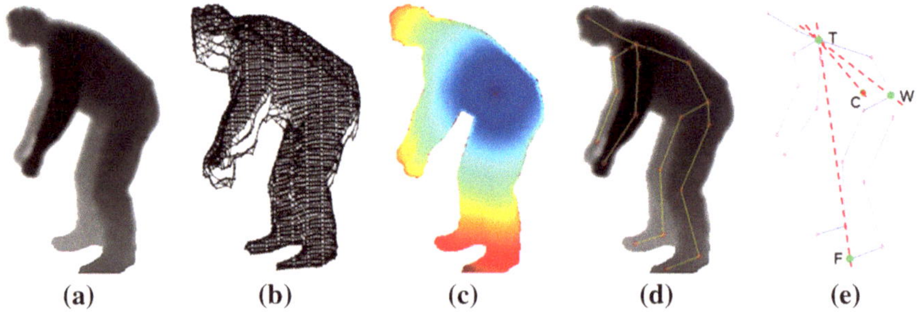

Fig. 15 Topological feature extraction approach. **a**) Original range image. **b**) Connected mesh computed starting from the 3D point cloud. **c**) Geodesic distance map. **d**) Reeb graph-based skeleton superimposed to the original range image. **e**) Line segments measured on skeleton during the feature extraction

node) named T and W in the Reeb graph shown in Fig. 15e, and the graph node having the minor height indicated as E in the same Fig. 15e. Hence, the topological feature vector is defined as follows:

$$\upsilon_T = \left(h_C, \angle\overline{TW}, \angle\overline{TC}, \angle\overline{TF}\right) \tag{25}$$

where h_C is the centroid's height with respect the floor plane and $\angle\overline{PQ}$ is the angle of 3D line segment $\overline{PQ}$ with respect the floor plane.

4.3 The Volumetric Approach

In order to discuss the volumetric-based feature extraction process, the 3D point cloud U computed by the range camera is defined as follows:

$$U = \left\{P_i = (X_i, Y_i, Z_i) \in R^3 | i = 1, \ldots, M\right\} \tag{26}$$

where X_i, Y_i, Z_i are the 3D world coordinates of the point P_i. The volumetric features exploit global information included into the 3D point cloud by considering two 3D cylindrical volumes V_{UP} and V_{DW} of radius R_i, as shown in Fig. 16, centered on the centroid C of the point cloud U and having world coordinates $C = (X_C, Y_C, h_C)$ in which $h_C > 0$ is the centroid height with respect the floor plane. Given the following subdivision of the cylinder's ray in regular intervals, $\forall k \in \{0, \ldots, N\}$:

$$R_0 = \min_{i \in \{1,\ldots,M\}} \|(X_i, Y_i) - (X_C, Y_C)\|, \; R_N = \max_{i \in \{1,\ldots,M\}} \|(X_i, Y_i) - (X_C, Y_C)\|, \; R_k = R_0 + k\frac{R_N - R_0}{N} \tag{27}$$

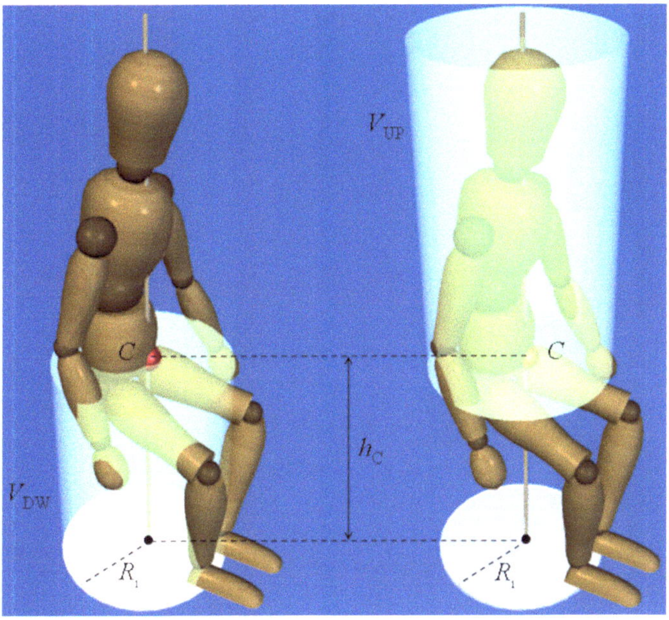

Fig. 16 3D cylindrical volumes used during the volumetric feature extraction

the total amount of points included in each cylinder is given by the following functions:

$$F_{UP}(k) = |\{i \in \{1,\ldots,M\} | \|(X_i, Y_i) - (X_C, Y_C)\| \leq R_k \wedge Z_i > h_C\}| \quad (28)$$

$$F_{DW}(k) = |\{i \in \{1,\ldots,M\} | \|(X_i, Y_i) - (X_C, Y_C)\| \leq R_k \wedge Z_i < h_C\}| \quad (29)$$

where $|\cdot|$ denotes the cardinality of a set. The volumetric feature vector can now be defined as follows:

$$\upsilon_V = \left(h_c,\ F_{UP}(N) - F_{DW}(N) \max_{1 \leq k < N} \Delta F_{UP}(k),\ \max_{1 \leq k < N} \Delta F_{DW}(k) \right) \quad (30)$$

where the operator Δ is the discrete derivative defined as follows:

$$\Delta F(k) \doteq F(k+1) - F(k) \quad (31)$$

In Fig. 17 the two functions defined by Eqs. 28 and 29 are plotted in correspondence of a 3D point cloud sampled for each main posture. The feature vector defined by Eq. 30 allows to keep very low the computational complexity of the feature extraction process although it is sufficient to discriminate reliably the four main postures since the spatial distribution of the 3D point cloud is dependent on the particular posture. Moreover, the computational simplicity is paid in terms of achievable level of detail in posture discrimination. Indeed, the feature vector υ_V is

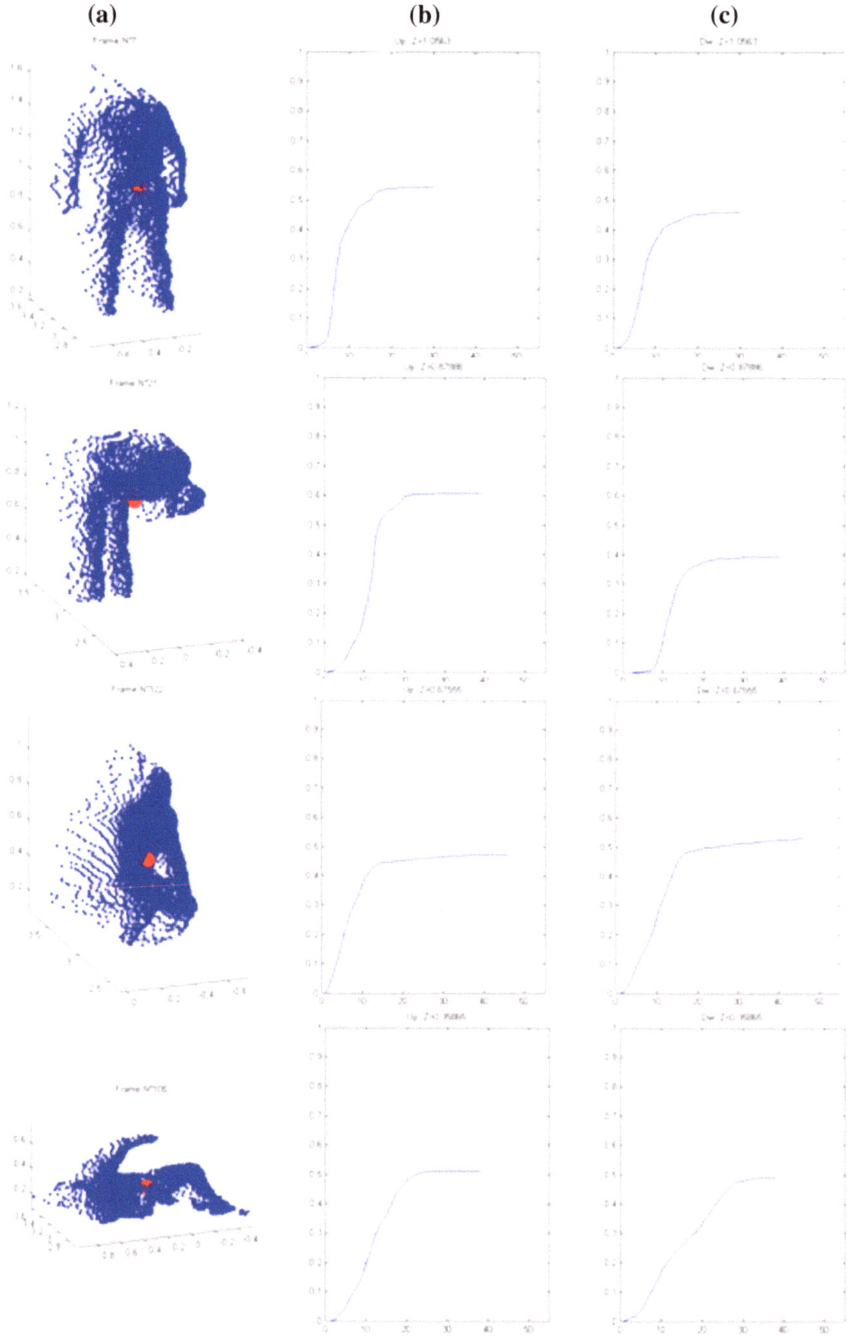

Fig. 17 Plotting of cylindrical volume cardinalities at the varying of ray values. **a**) 3D point clouds of the four main postures (from up to bottom): Standing, Bent, Sitting, Lying. **b**) Plots of the FUP function in correspondence of each posture. **c**) Plots of the FDW function in correspondence of each posture

unable to account for the position of body's segments like the υ_T feature vector does. However, the choice of the feature vector depends on the specific AAL application. If the positions of arms and legs are relevant (e.g., during the monitoring of rehabilitation exercises) then the topological approach is necessary; if it is needed to detect ADLs then the volumetric approach is sufficient (for fall-detection can be sufficient even the only h_C as it is discussed in the previous section).

4.4 Experimental Results

A good generalization ability during classification is definitely relevant since postures are not perfectly repeatable, the acquisition viewpoint varies in function of subject's position and some level of variation in range data is expected due to noise effects. This motivated the choice of a multi-class SVM classifier in conjunction with affine/isometric invariant features in order to discriminate the four main postures. Based on the principle of risk minimization, Support Vector Machines (SVMs) outperform other classifiers in terms of good performance in resolving non-linear and high dimensional problems with limited samples (high generalization ability). Moreover, SVMs try to find discriminative hyper-planes that maximize the margin between the classes overcoming, in a more natural way, the problem of over-fitting [54, 55]. The binary nature of SVM is adapted to the multi-class nature of the posture classification problem by using a one-against-one strategy. Since good results are documented in scientific literature related to posture recognition, a Radial Basis Function (RBF) kernel is used [56] and the associated parameters, namely regularization constant K and the kernel argument γ, are tuned according to a grid search procedure.

The best classification rates is experimentally found with the optimal parameters $(K;\gamma) = (1;32)$ for the topological approach and $(K;\gamma) = (1;64)$ for the volumetric one. A large dataset of 1200 samples, 300 for each posture, is collected in order to evaluate the classification performance. Postures are taken at various distances from the camera, ranging from 2.5 to 5 m. Confusion matrices are reported in Fig. 18 for both topological and volumetric features at distances of 2.5 and 5 m, whereas classification rates are reported in Fig. 19 for all intermediate distance values. As it is shown by reported results, topological features exhibit the best classification rate up to 3 m, whereas for distances greater than 3 m results are comparable with those of volumetric features. The training phase is done by using 200 samples, 50 samples for each posture, taken from only one viewpoint (the frontal view) in order to evaluate the generalization performance of the classification. Instead, the test phase is done by using the remaining 1000 samples taken from various viewpoints turning around the subject. The good generalization performance can be inferred by the reduced number of training samples adopted as support vectors, indeed about 40 % of the training set is used as support vectors. Results show the suggested features, both topological and volumetric, are suitable

(a)

	ST	BE	SI	LY
ST	1.000	0.002	0.000	0.000
BE	0.000	0.993	0.013	0.000
SI	0.000	0.005	0.983	0.006
LY	0.000	0.000	0.004	0.994

(b)

	ST	BE	SI	LY
ST	0.972	0.011	0.000	0.000
BE	0.019	0.967	0.026	0.000
SI	0.009	0.022	0.966	0.013
LY	0.000	0.000	0.009	0.987

(c)

	ST	BE	SI	LY
ST	0.988	0.009	0.000	0.000
BE	0.007	0.976	0.015	0.002
SI	0.005	0.014	0.982	0.009
LY	0.000	0.000	0.004	0.989

(d)

	ST	BE	SI	LY
ST	0.970	0.011	0.000	0.000
BE	0.018	0.972	0.025	0.003
SI	0.012	0.017	0.969	0.011
LY	0.000	0.000	0.006	0.986

Fig. 18 Confusion matrices for topological features **a**) at 2.5 meters and **c**) 5 meters. Confusion matri-ces for volumetric features **b**) at 2.5 meters and **d**) 5 meters

to exploit the full potential of range imaging most notably if used in conjunction with a classifier having good generalization capabilities (like SVM). Both topological and volumetric approaches (feature extraction and classification module) have been implemented on the embedded PC in c/c++ language achieving execution speed compliant with monitoring and surveillance purpose. When topological features were used the system worked at 5 fps with 87 % of execution time devoted to the feature extraction process. Instead, the system worked at 15 fps by using the volumetric features with an execution time of 60 % taken by the feature extraction process.

5 Discussion and Conclusion

The usage of TOF vision allows to solve some of the classic issues in background modeling and people segmentation, since depth information are not sensitive to illumination or shadows and can be used to detect more easily occlusions by exploiting the depth gap between people and occluding objects. Results shows the goodness of the proposed methods in real-time implementation and real AAL applications. The TOF camera experimented in this study is a state-of-the-art technology characterized by a very low noise and medium pixel resolution. Moreover, in order to keep this study as more general as possible, during data collection the TOF camera was set to a low integration time of 6 ms achieving so a noise level comparable with that of cheap cameras. The used camera is very compact and exploiting the proposed self-calibration algorithm it

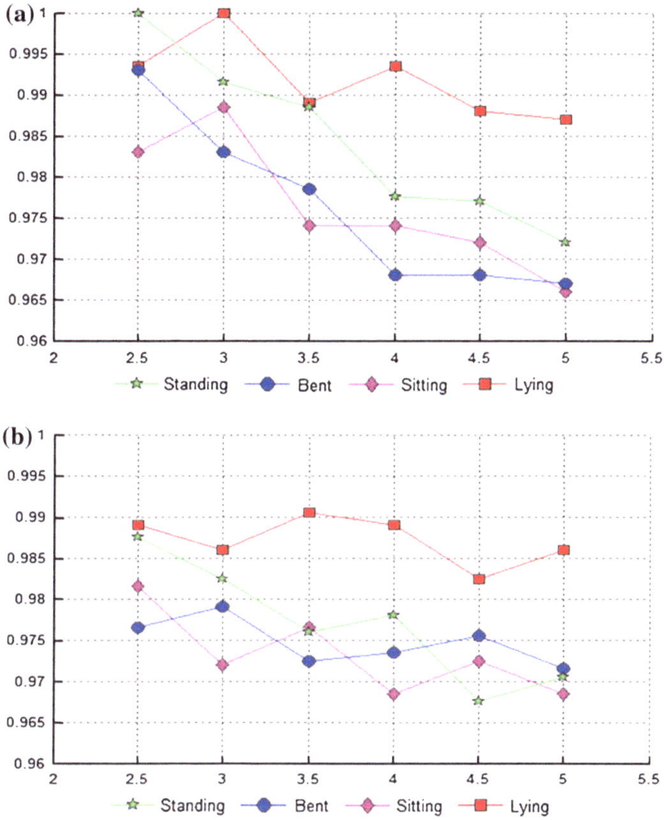

Fig. 19 Classification rates at varying of camera distance from 2.5 to 5 meters for both **a**) topological and **b**) volumetric features

can be installed simply without particular requirements or constraints. The suggested self-calibration procedure proved to be well suited for AAL application since it allowed to calibrate camera effectively without requirement of special calibration objects or user intervention but using only an automatic detection of floor plane that appeared to be always sufficiently visible (greater than 60 %) during falls recording in real dwelling rooms such as living room, kitchen, bed room, corridor and bathroom. The performance of the self-calibration algorithm was related to the amount of 3D points of the scene belonging to the floor plane. Better calibration estimations was obtained when at least the 30 % of the captured points belonged to the floor plane without floor covering and 40 % with carpet covering.

The presented fall-detector exploits the centroid height trend in order to detect fall events, thus the measurement precision is more important than accuracy. In other words, the systematic error does not affect fall-detection performance. The maximum estimated uncertainty in height measurement was less than 1.47 cm when camera was calibrated with floor plane variously covered by carpet and when at least 40 % of the captured points belonging to the floor plane. Moreover, between the TH1 threshold and the largest centroid peak value (35 cm) among all fall events was a difference of 5 cm (see dashed line in Fig. 11) that was definitely grater than the maximum uncertainty of 1.47 cm due to self-calibration procedure. The effect of the height measurement uncertainty on TH2 threshold was also evaluated. Taking into account all critical phase durations recorded during falls, the uncertainty in height measurement of 1.47 cm leaded to an uncertainty in the critical phase duration measurement less than 26 ms that was lower than the difference between the threshold TH2 and the maximum critical phase duration (860 ms) recorded during falls that was of 40 ms (see dashed line in Fig. 12). Since the third threshold TH3 was set to a very large time duration (at least of 4 s), the achieved precision did not play a critical role even in this case. Thus, the proposed threshold levels TH1, TH2 and TH3 provided a sufficient margin for successful detection of falls with respect to the height measurement precision. The three threshold in conjunction were able to detect all simulated falls without misdetection of ADLs as falls and vice versa, providing a 100 % of sensitivity and 100 % of specificity when occluding objects did not obstruct the camera's view. The tracking prediction mechanism allowed to estimate correctly the centroid height trend during simulated falls when some silhouette's portion was visible during critical and post fall phases (refer to Fig. 10). Fully occluded person movements during critical phase or during post fall phase gave rise to misdetections due to the impossibility to distinguish between a fall and a voluntary "lying down on floor and then stand up" (LYF) (critical phase occluded and post fall phase visible) or between a fall and a "fall with recovery" (FBWR or FFWR) (critical phase visible and post fall phase occluded). Instead, partial occlusions are correctly detected by evaluating the distance of the lower part of the segmented silhouette from the floor plane according to Eq. 11, allowing to adjust the position estimated by the particle filter. Thus, previously said misdetections demoted performance in presence of occluding objects leading to 97.3 % specificity and 80.0 % sensitivity. Segmentation and classification activities require about 25 ms per frame that seems reasonable since the minimum duration of the critical phase is about of 500 ms as indicate by Noury et al. [22]. Thus, a frame rate of 8 fps is fast enough for fall-detection purpose leaving available processing resources to be located for multiple 3D camera monitoring in order to deal with occlusions and limited FoV. It could be considered a limitation of presented studies that the trend of person's centroid height was determined from simulated falls in subjects with height greater than 1.55 m. However, the TH1 threshold not should be an issue for persons with height lower than 1.55 m, since they have a centroid height on average lower than taller persons. The TH2 threshold measured the critical phase duration. For persons with height lower than 1.55 m the critical phase duration could be shorter than that for person with higher height, thus TH2 should work correctly even for lower height. The TH3 threshold not should be an issue in any case since it works when the centroid height stays below the TH2 threshold. Results have been shown the feasibility to detect falls by using TOF camera highlighting related strength and weakness. The proposed fall-detector shows good performance also compared with other studies. In absence of occlusions performance is very similar to fall-detection system proposed by Bourke and Lyons [57] based on a bi-axial gyroscope sensor.

Other than for detection of falls, the capabilities of active vision have been demonstrated also for posture recognition in AAL contexts. Two feature extraction approaches, topological and volumetric, for the classification of four main postures (standing, bent, sitting and lying down) have been presented. The discrimination capabilities of the two feature extraction approaches are evaluated by using a machine learning approach and compared on the basis of a common dataset of simulated postures during simple ADLs. The different discrimination capabilities and execution speeds offered by the two approaches allow to satisfy different requirements exhibited by AAL applications. In fact, gathered posture details and operational distance from the camera are usually inversely proportional. For instance, rehabilitation exercises can be performed at few meters from the camera (e.g., less than 3 m) and many postural details are required in order to check the correctness of exercise execution, whereas critical events can occur at a greater distance from the camera (more than 3 m) but few postural detail are usually sufficient for detection of critical events. The topological features describe the human posture at a high level of detail exploiting the full potential offered by range imaging: many body segments can be discriminates such as head, trunk, arms and legs. As it is shown by reported results, topological features exhibit the best classification rate up to 3 m, whereas for distances greater than 3 m results are comparable with those of volumetric features. However, the high level of postural detail achieved with the topological features is paid in terms of computational workload (up to 5 fps). Volumetric features reflecting the spatial distribution of 3D point cloud provide a lower level of detail in posture discrimination, but they have the advantage to be less computationally expensive (up to 15 fps). The choice for one or the other depends on the specific AAL application. The results suggest high accuracy of topological features at distances up to 3 m, whereas beyond volumetric and topological approaches give similar classification performance (greater than 96.5 % in both cases).

References

1. N. Foldi, L. Kaplan, J. Ly, O. Nikelshpur, M. Lucy, B. Jeffrey, ADL functions and relationship to cognitive status. Alzheimer's Dement. J. Alzheimer's Assoc. **7**(4), S244 (2011)
2. N. Shah, M. Kapuria, K. Newman, in *Embedded activity monitoring methods*. Activity Recognition in Pervasive Intelligent Environments, vol 4(13) (Atlantis Press, Amsterdam, 2011), pp. 291–311
3. B. Kröse, T. Oosterhout, T. Kasteren, *Activity Monitoring Systems in Health Care*, in Computer Analysis of Human Behavior, vol. 12 (Springer, London, 2011), pp. 325–346

4. A. Agarwal, B. Triggs, Recovering 3D human pose from monocular images. IEEE Trans. PAMI **28**(1), 44–58 (2006)
5. A. Fossati, M. Dimitrijevic, V. Lepetit, P. Fua, From canonical poses to 3D motion capture using a single camera. IEEE Trans. PAMI **32**(7), 1165–1181 (2010)
6. S.N. Lim, A. Mittal, L.S. Davis, N. Paragios, Fast illumination invariant background subtraction using two views: error analysis, sensor placement and applications. Proc. Comput. Vis. Pattern Recogn. **1**, 1071–1078 (2005)
7. R.I. Hartley, A. Zisserman, *Multiple View Geometry in Computer Vision*, 2nd ed. (Cambridge University Press, Cambridge, 2004)
8. V. Ganapathi, C. Plagemann, S. Thrun, D. Koller, Real time motion capture using a single Time-Of-Flight camera, in *Proceedings of CVPR*, pp. 755–762 2010
9. W. Li, Z. Zhang, Z. Liu, Action recognition based on a bag of 3D points, in *Proceedings of CVPRW*, pp. 9–14 2010
10. C. Plagemann, V. Ganapathi, D. Koller, S. Thrun, Real-time identification and localization of body parts from depth images, in *Proceedings of ICRA*, pp. 3108–3113 2010
11. S. Oprisescu, C. Burlacu, V. Buzuloiu, Action recognition using time of flight cameras, in *Proceedings of COMM*, pp. 153–156 2010
12. C. Rougier, E. Auvinet, J. Rousseau, M. Mignotte, J. Meunier, Fall detection from depth map video sequences. Lect. Notes Comput. Sci. **6719**(2011), 121–128 (2011)
13. D. Falie, M. Ichim, Sleep monitoring and sleep apnea event detection using a 3D camera, in *Proceedings of 8th IEEE International Conference on Communications (COMM)*, pp. 177–180 2010
14. M. Grassi, A. Lombardi, G. Rescio, M. Ferri, P. Malcovati, A. Leone, G. Diraco, P. Siciliano, M. Malfatti, L. Gonzo, An integrated system for people fall detection with data fusion capabilities based on 3D TOF camera and wireless accelerometer, in *Proceedings of IEEE Sensors*, pp. 1016–1019 2010
15. G. Diraco, A. Leone, P. Siciliano, Geodesic-based human posture analysis by using a single 3D TOF camera, in *Proceedings of IEEE International Symposium on Industrial Electronics (ISIE)*, pp. 1329–1334 2011
16. A. Leone, G. Diraco, P. Siciliano, Detecting falls with 3D range camera in ambient assisted living applications: a preliminary study. Med. Eng. Phys. J. **33**(6), 770–781 (2011)
17. www.xbox.com/Kinect
18. A. Kolb, E. Barth, R. Koch, TOF-sensors: new dimensions for realism and interactivity, in *Proceedings of Computer Vision and Pattern Recognition Workshops*, pp. 1–6 2008
19. S. Foix, G. Alenyà, C. Torras, Lock-in Time-Of-Flight (TOF) cameras: a survey. IEEE Sens. J. **11**(9), 1917–1926 (2011)
20. S.A. Guomundsson, H. Aanaes, R. Larsen, Environmental effects on measurement uncertainties of Time-Of-Flight cameras. Proc. IEEE ISSCS **1**, 1–4 (2007)
21. D. Lee, Effective Gaussian mixture learning for video background subtraction. IEEE Trans. Pattern Anal. Mach. Intell. **27**(5), 827–832 (2005)
22. N. Noury, P. Rumeau, A.K. Bourke, G. ÓLaighin, J.E. Lundy, A proposal for the classification and evaluation of fall detectors. J. IRBM **29**(6), 340–349 (2008)
23. M.C. Chung, K.J. McKee, C. Austin, H. Barkby, H. Brown, S. Cash, J. Ellingford, L. Hanger, T. Pais, Posttraumatic stress disorder in older people after a fall. Int. J. Geriatr. Psychiatry **24**(9), 955–964 (2009)
24. S. Sadigh, A. Reimers, R. Andersson, L. Laflamme, Falls and fall-related injuries among the elderly: a survey of residential-care facilities in a Swedish municipality. J. Commun. Health **29**, 129–140 (2004)
25. A. Shumway-Cook, M.A. Ciol, J. Hoffman, B.J. Dudgeon, K. Yorkston, L. Chan, Falls in the medicare population: incidence, associated factors, and impact on health care. J. Phys. Ther. **89**(4), 324–332 (2009)
26. S.R. Lord, C. Sherrington, H.B. Menz, *Falls in Older People. Risk Factors and Strategies for Prevention* (Cambridge University Press, Cambridge, 2007)

27. S. Elliott, J. Painter, S. Hudson, Living alone and fall risk factors in community-dwelling middle age and older adults. J. Commun. Health **34**, 301–310 (2009)
28. M. Shaou-Gang, S. Fu-Chiau, H. Chia-Yuan, A smart vision-based human fall detection system for telehealth applications, in *Proceedings of the 3rd Telehealth Conference*, pp. 7–12 2007
29. B. Jansen, R. Deklerck, Context aware inactivity recognition for visual fall detection, in *Proceedings of IEEE Pervasive Health Conference*, pp. 1–4 2006
30. R. Cucchiara, A. Prati, R. Vezzani, A multi-camera vision system for fall detection and alarm generation. Expert Syst. J. **24**(5), 334–345 (2007)
31. http://www.mesa-imaging.ch
32. M.A. Fischler, R.C. Bolles, Random sample consensus: a paradigm for model fitting with applications to image analysis and automated cartography. Commun. ACM **24**(6), 381–395 (1981)
33. http://www.xsens.com
34. C. Fabien, B. Deepayan, A. Charith, S.H. Mark, Video based technology for ambient assisted living: a review of the literature. Environments **3**, 253–269 (2011). IOS Press
35. D.S. Lee, Effective Gaussian mixture learning for video background subtraction. IEEE Trans. Pattern Anal. Mach. Intell. **27**(5), 827–832 (2005)
36. C. Stauffer, W. Grimson, Adaptive background mixture models for real-time tracking, in *Proceedings of IEEE Computer Vision and Pattern Recognition Conference*, pp. 246–252 1999
37. http://sourceforge.net/projects/opencvlibrary
38. R.E. Kalman, A new approach to linear filtering and prediction problems. Trans. ASME J. Basic Eng. **82**(Series D), 35–45 (1960)
39. M. Isard, A. Blake, CONDENSATION—conditional density propagation for visual tracking. Int. J. Comput. Vis. **29**(1), 5–28 (1998)
40. N. Noury, A. Fleury, P. Rumeau, A.K. Bourke, G.O. Laighin, V. Rialle, J.E. Lundy, Fall detection—principles and methods, in *Proceedings of the 29th IEEE EMBS*, pp. 1663–1666 2007
41. J. Parkkari, P. Kannus, M. Palvanen, A. Natri, J. Vainio, H. Aho et al., Majority of hip fractures occur as a result of a fall and impact on the greater trochanter of the femur: a prospective controlled hip fracture study with 206 consecutive patients. Calcif. Tissue Int. **65**, 183–187 (1999)
42. T.B. Moeslund, A. Hilton, V. Kruger, A survey of advances in vision-based human motion capture and analysis. J. CVIU **104**(2–3), 90–126 (2006)
43. J. Deutscher, I. Reid, Articulated body motion capture by stochastic search. Int. J. Comput. Vis. **61**(2), 185–205 (2005)
44. G. Mori, J. Malik, Recovering 3D human body configurations using shape contexts. IEEE Trans. PAMI **28**(7), 1052–1061 (2006)
45. R. Navaratnam, A.W. Fitzgibbon, R. Cipolla, Semi-supervised learning of joint density models for human pose estimation, in *Proceedings of BMVC*, vol. 2, pp. 679–688 2006
46. K. Mikolajczyk, D. Schmid, A. Zisserman, Human detection based on a probabilistic assembly of robust part detectors, in *Proceedings of European Conference on Computer Vision (ECCV)*, pp. 69–81 2004
47. C. Sminchisescu, A. Kanaujia, Z. Li, D. Metaxas, Discriminative density propagation for 3D human motion estimation. Proc. Comput. Vis. Pattern Recogn. **1**, 390–397 (2005)
48. L. Ren, G. Shakhnarovich, J.K. Hodgins, H. Pfister, P.A. Viola, Learning silhouette features for control of human motion. J. ACM TOG **24**(4), 1303–1331 (2005)
49. Q. Delamarre, O. Faugeras, 3D articulated models and multi view tracking with physical forces. J. CVIU **81**(3), 328–357 (2001)
50. A. Leone, G. Diraco, P. Siciliano, Topological and volumetric posture recognition with active vision sensor in AAL contexts, in *Proceedings of 4th IEEE International Workshop on Advances in Sensors and Interfaces (IWASI)*, pp. 110–114 2011

51. N. Werghi, Y. Xiao, J.P. Siebert, A functional-based segmentation of human body scans in arbitrary postures. J. T-SMCB **36**(1), 153–165 (2006)
52. G. Reeb, Sur les points singuliers d'une forme de Pfaff complétement intégrable ou d'une fonction numérique. C.R. Acad. Sci. Paris **222**, 847–849 (1946)
53. A. Verroust, F. Lazarus, Extracting skeletal curves from 3-D scattered data. Vis. Comput. **16**(1), 15–25 (2000)
54. C.J.C. Burges, A tutorial on support vector machines for pattern recognition. Data Min. Knowl. Disc. **2**(2), 121–167 (1998)
55. I. Steinwart, A. Christmann, *Support vector machines* (Springer, New York, 2008)
56. H. Dongcheol, C. Wallraven, L. Seong-Whan, View invariant body pose estimation based on biased manifold learning, in *Proceedings of ICPR*, pp. 3866–3869 2010
57. A.K. Bourke, G.M. Lyons, A threshold-based fall-detection algorithm using a bi-axial gyroscope sensor. Med. Eng. Phys. J. **30**(1), 84–90 (2008)